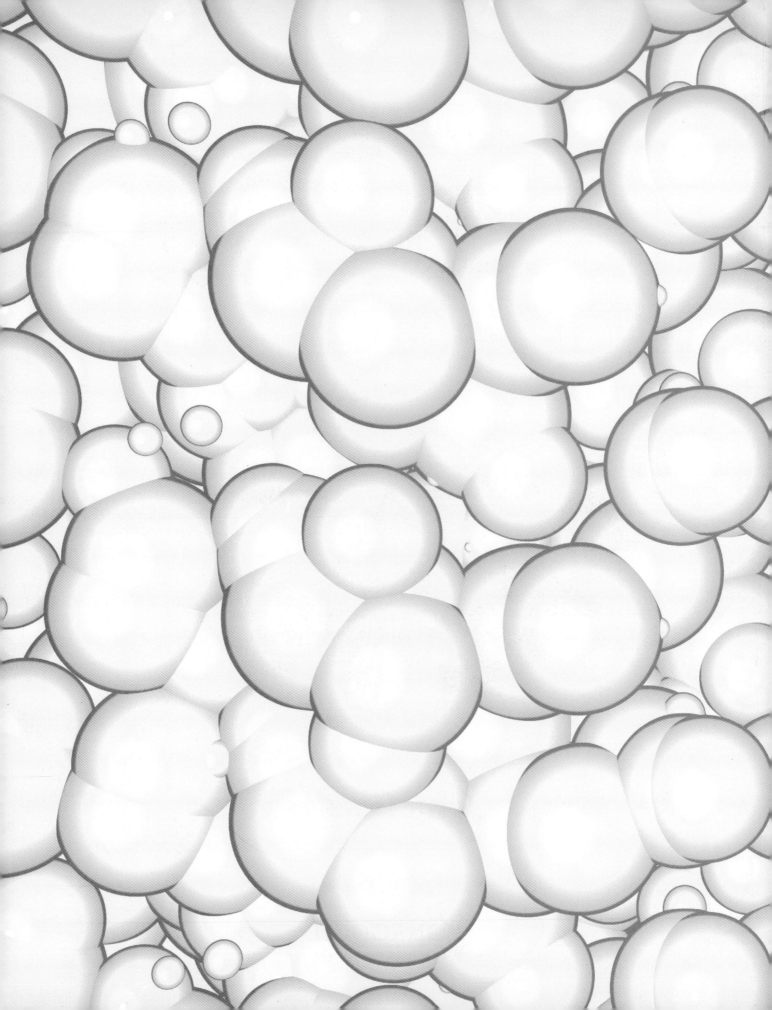

TRUE FACTS

John Guest

Bath · New York · Singapore · Hong Kong · Cologne · Delhi · Melbourne

Author: John Guest
Consultants: Ben Robinson, Ade Scott-Colson,
Tony Sizer and John Williams
This edition produced by Tall Tree Ltd, London

First published by Parragon in 2007

Parragon
Queen Street House
4 Queen Street
Bath BA1 1HE, UK

ISBN 978-1-4054-9546-2

Printed in Malaysia

Contents

Introduction

Discover all that's weird and wonderful in this amazing book. Find out about the greatest, maddest, and nastiest people who've ever lived. Take a look at some of the odder parts of life and the crazy ideas and awesome technology humans have come up with. But humans aren't the whole story. There are creatures much stranger than us and parts of our planet where things happen that you just wouldn't believe—and once you get into outer space things start to get very weird indeed...

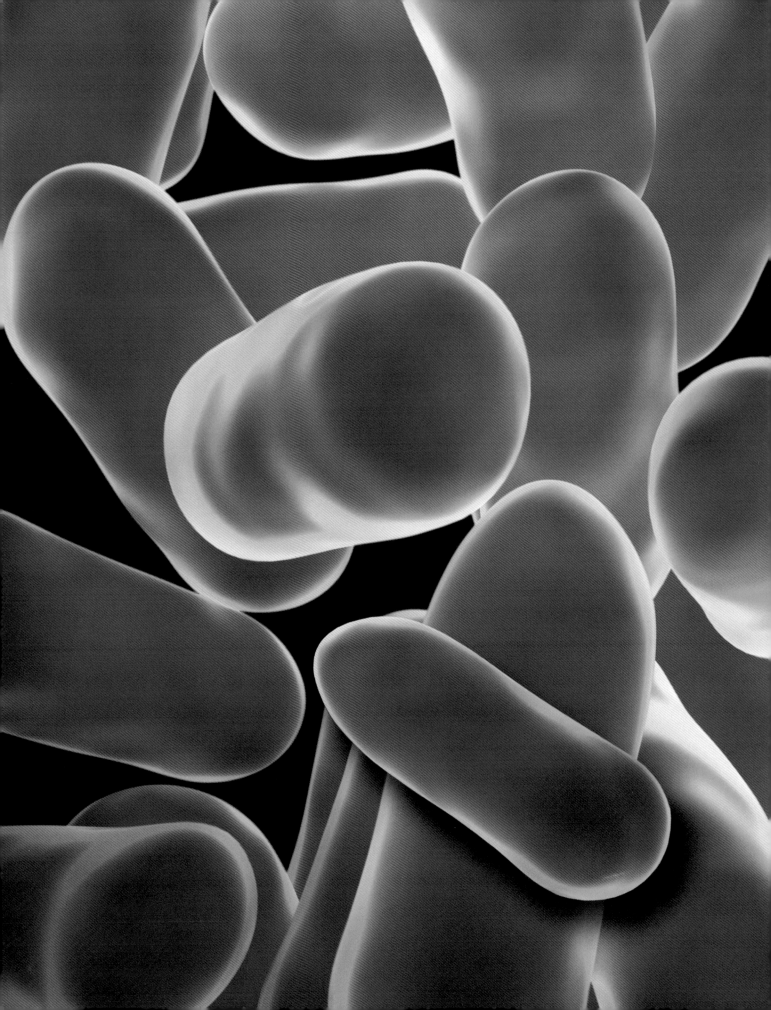

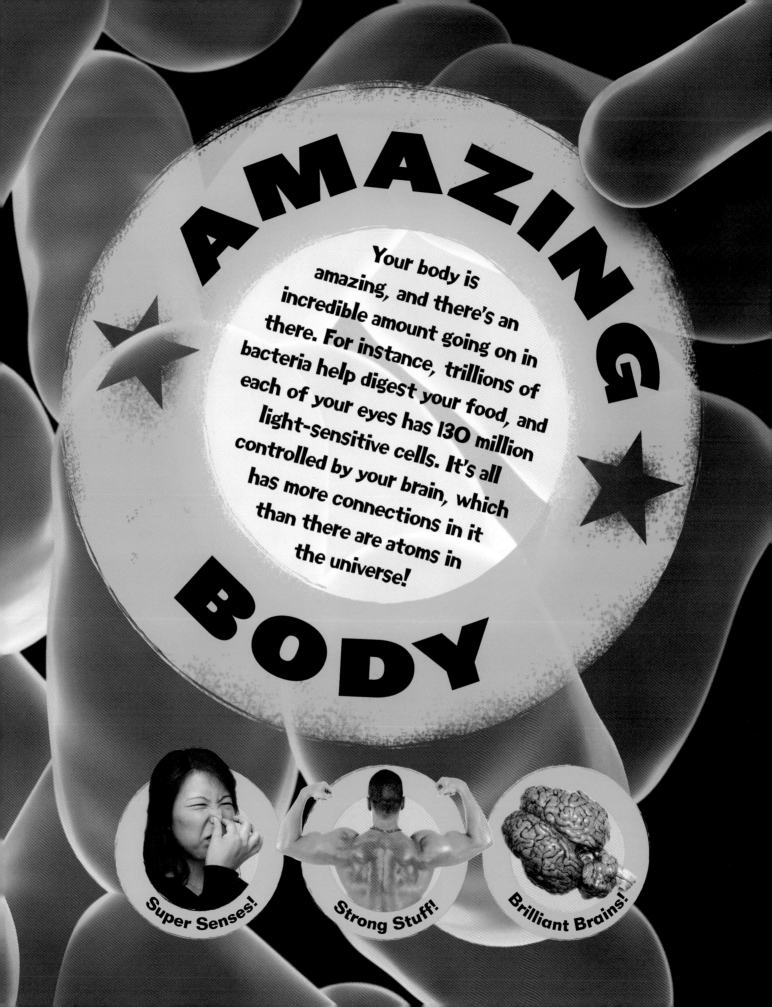

AMAZING BODY

Your body is amazing, and there's an incredible amount going on in there. For instance, trillions of bacteria help digest your food, and each of your eyes has 130 million light-sensitive cells. It's all controlled by your brain, which has more connections in it than there are atoms in the universe!

Super Senses!

Strong Stuff!

Brilliant Brains!

Brilliant Brains

It looks like a pile of fat spaghetti, but your brain is actually one of the most amazing things in the universe.

☛ There are more connections in your brain than there are atoms in the universe.

☛ Everyone has their own unique pattern of wrinkles on their cerebral cortex.

If you're a woman, your brain weighs 2.5 percent of your body weight, on average, while if you're a man, it weighs 2 percent – but men's brains are heavier on average because men are bigger than women.

Brain Food

✎ Your brain is 80 percent water.

✎ One-fifth of all the oxygen that you breathe goes to your brain.

✎ Your brain demands 25 percent of your blood supply.

✎ About 29 ounces of blood shoot through your brain every minute.

✎ Scientists can now grow human brain cells in a laboratory dish.

Surgeons can operate on a brain while the person is conscious, because the brain doesn't feel pain when it is cut.

TRUE NERVE FACTS

☞ Neurons, or nerves, carry messages around your body using electricity.

☞ If you strung all your nerves end to end, they would measure 50 miles.

☞ Nerves can carry signals at a speed of 100 yards a second.

☞ When you were in the womb, you grew 250,000 nerves every second.

☞ After you're born, very few new nerves are made. By the time you're ten, half of them have died off. But don't worry, you've still got plenty left.

➤ Your brain has about 100,000,000,000 (100 billion) neurons.

➤ Each neuron is connected to about 25,000 other neurons.

Super Senses

Your body tells you what's going on using your nose, ears, eyes, tongue, and sense of touch.

In 2001, the U.S. army did some research to develop a stinkbomb to deal with rioters—and they found that the most devastating smell was poop mixed with rotting onions.

Super Smell

➡➡ There are 1,000 different kinds of stink receptor in your nose.

➡➡ Your nose can identify an infinite number of aromas.

➡➡ The smell of a raspberry comes from 300 different chemicals.

Super Sight

☛ Each of your eyes has 130 million light-sensitive cells built into an area the size of a postage stamp.

☛ You have over 5 million color-detecting cones in the retina of each eye.

☛ Your eyes blink 10–24 times a minute—that's 415 million times in a lifetime.

☛ You have 200 eyelashes on each eye to protect them from dust.

☛ Each of your eyeballs weighs about 1 ounce and is full of water.

Super Hearing

✎ Your ears can tell the difference between two sounds even if they are only a ten-millionth of a second apart.

✎ Your ears can hear sounds as high-pitched as 20,000 hertz and as low as 20 hertz.

✎ Dogs can hear from 50,000 hertz down to 15 hertz. Dolphins can hear up to 120,000 hertz.

✎ Your ears go pop when you dive or go up in a plane because your eardrums move in and out to make sure there is equal pressure inside and outside. Otherwise your eardrums might burst.

Super Touch

✎ Your fingers are so sensitive they can feel an object move even if it only moves a thousandth of an inch.

✎ There are 200,000 hot and cold receptors in your skin, 500,000 touch and pressure receptors, and nearly 3 million pain receptors.

Super Taste

➤ Your tongue has 8,000 taste buds.

➤ Your tongue can taste only five different tastes—sweet, sour, salty, bitter, and savory.

➤ The savory taste is often called umami, a Japanese word.

➤ A tongue can taste one drop of lemon juice in 129,000 drops of water.

Flaky Skin

The biggest organ in your body, your skin keeps the outside out and the insides in!

Skin-deep

☛ Skin is the heaviest part of your body and weighs about 5–9 pounds.

☛ Taken off and laid flat, your skin would be between 15 and 20 feet square.

☛ Your skin is $1/4$ inch thick on the soles of your feet and just $1/50$ inch thick on your eyelids.

Second Skin

✎ Skin cells fall off at a rate of 30,000–40,000 a minute, so every month your outer layer of skin is completely renewed.

✎ Each year you lose almost 2 pounds of skin—enough to fill a large cookie jar.

If you think your bedsheets are clean, think again. There are probably a million of these tiny dust mites (above) chomping away on your old skin flakes right now. Tasty!

☞ Zits form when glands that make a waxy, oily substance called sebum get overworked and clog up.

☞ Blackheads turn black because a substance called keratin (the stuff hair is made from) builds up in glands and pushes out the sebum so it is exposed to air.

☞ Whiteheads are white if there is only a small amount of sebum at the top of a gland and it is not exposed to the air.

☞ It used to be thought that zits could be caused by a diet high in fat or sugar. Now scientists think other factors, such as hormones, are more important.

Over half the dust in your home is dead skin cells. Dandruff is also made of dead skin, from your scalp, which has clumped together and looks like this close up.

➣ Once your blood clots, or thickens, 16 different chemicals work together to turn the wet clot into a dry scab.

➣ Scabs start to form less than 10 seconds after you cut yourself.

Growing Parts

There are parts of our bodies that keep on growing and we regularly have to cut. Well, most of us do...

TRUE HAIR FACTS

☞ Hair grows faster than anything else in your body.

☞ You've got about 100,000 hairs growing from your head.

☞ Fair-haired people have a lot more hairs than redheads— 150,000 versus 90,000.

☞ Hair grows about ½ inch a month.

➤➤ Hot weather makes your hair grow faster.

➤➤ Most hairs fall out once they get to about 35 inches long.

➤➤ A rope made from 1,000 hairs could lift a full-grown man.

➤➤ Your hairs stand on end when you're scared because muscles pull them up to make you look bigger.

➤➤ Humans have the same number of hairs on their bodies as chimpanzees.

Bushy Beards

☞ Norwegian Hans Langseth had a beard over 16 feet long when he died in 1927.

☞ Beards are the fastest growing hairs on the human body. If the average man never trimmed his beard, it would grow to nearly 30 feet long in his lifetime.

☞ Beards, and all other hair, are made from the tough substance keratin. Nails, hooves, and feathers are also made from keratin.

Nail Bites

✎ A fingernail takes about six months to grow from base to tip.

✎ Your middle fingernail grows fastest, your thumbnail grows slowest.

✎ Your fingernails would grow 92 feet long in your life if you didn't cut them.

✎ The longest ever nails belonged to Sridhar Chillal of Pune in India, whose nails were a combined length of over 23 feet when he finally decided to cut them off— and sell them for $200,000 on the Internet!

✎ People have painted their nails different colors using nail polish for at least the last 5,000 years.

Bodily Fluids

Your body produces a lot of liquid, and sometimes it leaks out.

➤➤ All the breathing tubes inside your body are lined with sticky snot called mucus.

➤➤ Snot moves through your breathing tubes at $1/2$ inch an hour.

➤➤ You recycle about 2 pints of snot a day by swallowing it.

➤➤ Snot is not just useless goo. It's the stuff your body oozes out to trap dirt and protect the lining of your airways.

Cows can make 200 times as much saliva as humans. Cow saliva contains an antibiotic that may one day be used as a medicine.

Boogers

☞ Boogers are dried-up snot—that is, mucus without the water. But they're also packed with all kinds of debris you've breathed in, including dust, pollen, germs, sand, fungi, and smoke.

☞ Your boogers take on the color of the dust in the places where you've been.

Tears wet your eyes each time you blink—about 25 times a minute. It takes 100 tears to fill a teaspoon. You cry 14 gallons in a lifetime, that's about 1,850,000 tears.

TRUE SNEEZE FACTS

☛ Sneezing is a reflex—which means it happens automatically. Part of the sneeze reflex is closing your eyes.

☛ Sneezing throws your head back with more force than a roller coaster does.

☛ People probably say "Bless you" when you sneeze because in medieval times it was thought to be a sign of the plague, so it meant you might be about to die.

☛ Photic sneezing refers to sneezing when you see a bright light. One in four people are affected by it.

☛ The world sneezing record is held by 12-year-old British girl Donna Griffiths, who sneezed 1 million times in a row!

➤ Your ears have 2,000 glands for making ear wax.

➤ Ear wax varies. It can be gray, yellow, orange, or brown. It can be moist or dry.

➤ Your ears make new wax all the time. Wax is dropping out of your ears right now.

When you sneeze, air and snot hurtle out of your nose at 100 mph.

Strong Stuff

Your muscles are the parts of you that hold you together and get you moving.

Muscles make up over 40 percent of your body's entire weight. If all the muscles in your body pulled together, they could lift a bus.

A smile is said to use 17 muscles, while a frown uses 43. In your face, there are a lot of small muscles, which you use to make expressions.

Your body's biggest muscle is in your buttocks, the gluteus maximus. *Gluteus* is the Latin word for "buttocks." The gluteus maximus is the biggest of the three muscles in your buttocks.

➤➤ Your body's largest muscles are made from hundreds of bundles of muscle fiber.

➤➤ Your body's longest muscle is the sartorius on the inner thigh. It gets its name from an old word for tailor. Tailors sat cross-legged, and this is the muscle you use to cross your legs.

➤➤ Your body's widest muscle is the external oblique, which runs around the side of the upper body.

Twitchy Fibers

☛ Your skeletal muscles are made up from 6 billion thin strings called fibers.

☛ Each fiber is made from tiny threads called fibrils.

☛ Fibers are slow-twitch for endurance or fast-twitch for bursts of strength. Sprinters have a lot of fast-twitch fibers, but they would be useless in a marathon.

✎ You have more than 640 muscles on the outside of your body, called skeletal muscles.

✎ Skeletal muscles work in pairs to move your body around.

➤➤ Your tongue is one of the strongest muscles in your body. Lick that!

➤➤ The smallest muscle in your body is less than 2/25 inch long. It's called the stapedius and is in your inner ear.

➤➤ Men might be surprised to learn that women have the strongest muscle for its size. Weight-for-weight, a woman's uterus is stronger than any other muscle.

☛ The most forceful muscle in your body is your masseter, the muscle you chew with. The masseter can chew with a pressure of up to 365 pounds!

☛ Plant-eating animals, such as cows, have the strongest masseters. They need all the help they can get to munch up that tough grass.

TRUE JAW FACTS

You've Got Guts!

Look at where your food goes when you swallow—assuming it's not all down your front.

TRUE FOOD FACTS

☞ It takes eight seconds for food to get to your stomach.

☞ Your stomach takes about six to seven hours to process a three-course meal.

☞ Food then needs three to five hours to get through your small intestine.

☞ In total, it takes your gut about 24 hours to get undigested food right the way through and out the other end.

☞ Fat is so hard to digest that it congeals into large globs in your insides. It's finally dissolved away by a substance called bile, which oozes out of your gallbladder.

The average person manages to stuff down over 30 tons of food in their lifetime—that's the weight of about 80 horses, six elephants, or half a blue whale.

➤➤ You move food along inside you using waves of squeezing muscles. So food would go through even if you stood on your head.

➤➤ When empty, your stomach holds barely 2 cups, but after a big meal it can stretch to more than 3 quarts.

Your small intestine has a huge area for absorbing food because it is wrinkled up into tiny little knobbles called villi. Ironed flat, it would be big enough to carpet your bedroom.

Acid Attack

🖊 Your stomach produces acid strong enough to dissolve a lump of bone in a few hours.

🖊 The juices your stomach uses to digest food contain acid that is so strong it can even eat metal!

🖊 Your body makes more than 1½ gallons of digestive juices a day and recycles much of it.

Gut-tastic!

☛ If you could lay your intestines, or gut, out straight, they would be nearly six times as long as you are tall.

☛ Your small intestine is divided into your duodenum, your jejunum, and your ileum.

☛ Your large intestine is divided into your caecum, your colon, and your rectum.

Have a Heart

Blood brings nutrients to your body's cells and takes away their waste. It's pumped around your body by your powerful heart.

Pump Power

➤➤ During your life, your heart will probably pump about 44 million gallons of blood—that could fill a swimming pool deep enough to hold the Empire State Building.

➤➤ Your heart beats approximately 100,000 times a day—and about 2.5 billion times in your lifetime.

➤➤ Squeeze a tennis ball until you make a good dent. That's about how hard your heart muscles have to squeeze to pump blood!

➤➤ Your heart pumps with enough power during a single day to drive an average car 20 miles.

Most people's blood belongs to one of four groups: A, O, B, or AB. Blood is also either Rhesus + or Rhesus −, depending on whether it has a chemical in it that was first discovered in Rhesus monkeys.

TRUE BLOOD FACTS

☞ You have about 3–6 quarts of blood in your body, depending on how big you are.

☞ All your blood travels right around your body once every 20 seconds.

☞ In one day, your blood travels 12,000 miles—that's half way around the world.

☞ The largest blood vessel, called the aorta, is about the size of a garden hose. The aorta takes blood from the heart to smaller blood vessels.

☞ The smallest blood vessels, called capillaries, are about one tenth as thick as one of the hairs on your head.

➤➤ Blood is a mix of red and white cells and tiny pieces of cell called platelets, all floating in a yellowy liquid called plasma.

➤➤ Red blood cells look like tiny buttons and are responsible for carrying oxygen around the body.

White Blood Cells

☞ White cells are much bigger than red cells and are involved in the fight against infections.

☞ Your body makes white blood cells in the marrow in your bones. You make about a billion of them every day.

Your kidneys are filters that take harmful waste out of your blood. They turn this waste into pee and send it to your bladder ready for when you go to the bathroom.

Poop and Pee

As well as putting fuel into our bodies, we also need to get rid of stinky waste.

Perfect Pee

☛ When full, your bladder swells up like a balloon, and it is just about as thin-skinned.

☛ Your pee is mostly water, plus a bit of salt, and the chemical compound urea.

☛ Urea is rich in nitrogen, which crops need to grow. But don't go peeing in fields—farmers have their own supply of urea that they put in sprays.

What a Relief!

✏ You pee between one and two quarts of urine a day.

✏ In your life, you pee enough to fill a small swimming pool. That's about 10,000 gallons.

➻ Absolutely fresh pee has no smell. But as soon as it comes out of your body, the urea in it starts to break down and, boy, does it stink!

➻ The smell in old pee comes from the gas ammonia. Ammonia is the gas found in smelling salts, fertilizers, and explosives!

Toilet Tales

➤➤ The average human spends between six months and a year of their life on the toilet.

➤➤ The average adult goes to the bathroom about four times a day.

TRUE POOP FACTS

☛ A good poop is about 75 percent water. Diarrhea is even more watery. Water is absorbed as poop passes through the intestine, so the longer a poop stays inside, the drier it will be.

☛ Of the non-watery part of a poop, about one-third is dead bacteria—the victims of the battle inside to digest your food.

☛ Another one-third is stuff we can't digest, such as cellulose (fiber). This helps move feces along, because it gives the gut muscles something to grip onto.

☛ The final third is a slurpy mixture of fats, dead cells, salts, mucus, and live bacteria. Mmm!

☛ Poop is brown because it contains an iron-rich substance called bilirubin, which comes from old red blood cells.

☛ Poop with a lot of fat or gas in it floats.

Newborn babies' poop is green. That's because they couldn't poop inside the womb, so bile digestive juices have been building up for months with nowhere to go.

Vile Vomit and Bumptious Burps

Sometimes our bodies give off more
than we've bargained for.

Throwing up isn't easy. First your abdomen muscles squeeze down on your stomach. Pressure builds up on the mush in your stomach, a little at a time. Then suddenly a valve opens into the tube that leads to your mouth—and out comes the vomit.

Vomit

✎ When you're just about to throw up, you start to dribble. This protects your teeth from the strong stomach acids in vomit.

✎ You throw up when your brain detects poisons in your stomach or blood.

✎ There's a place in your brain called the vomit center, which tells you to lean forward and open your mouth just before you throw up. That saves a lot of mess.

✎ Mice, rabbits, rats, and horses are the only mammals that don't vomit.

VOMIT RECIPE
Mix some mushed-up, half-digested food from your stomach with slimy stomach mucus and nasty digestive chemicals, then add a dash of saliva. Mmmm, tasty!

Vomit is often a lovely color green because it contains a chemical used for digestion called bile. Bile comes from far down in your digestive system beyond your stomach.

The word 'nausea' comes from the Greek word for seasickness.

TRUE BURP FACTS

☞ Burps are the sound of gas roaring from your stomach up through your gullet.

☞ Most people burp about 10–15 times a day.

☞ Burping gets rid of a quart of gas from your stomach every day.

☞ Your stomach would inflate like a balloon if you didn't burp.

☞ A burp was recorded at 118.1 decibels—as loud as a car alarm.

Personal Zoo

There are trillions of tiny organisms called microbes living in your body!

Boy, Your Feet Smell!

Feet can stink because bacteria and fungi adore the warmth and moisture of socks and shoes—and all the sweat and dead skin cells in them. It's the bacteria that give off the blue-cheese aroma of some feet.

Skin is usually just about the least friendly place on your body for microbes, because it's very dry. There are only about 100 of them living on each square 1/6 inch.

> → More microbes live in your mouth than people live on Earth!

> → The microbes in there include hundreds of species of bacteria, fungi, protozoa, and viruses.

> → One bacterium, *Streptococcus mutans*, does all the harm— it's this that rots your teeth.

Love Your Guts

✎ Your guts are a real metropolis for bacteria. There are trillions of them down there.

✎ In fact, every gram of fluid taken from the large intestine scoops up 10 trillion bacteria!

✎ Bacteria in your guts make vitamin K, an essential vitamin your body needs.

✎ *Lactobacillus acidophilus* are nice bacteria that crowd out more unsavory bugs in your gut.

✎ Travelers sometimes swallow capsules of *Lactobacillus* to avoid diarrhea.

TRUE FART FACTS

☞ If it wasn't for the bacteria that live in your guts, your farts wouldn't smell.

☞ There's a crowd of bacteria in your guts called *Methanobacterium smithii*, which break down tough food and make methane gas.

☞ The methane and hydrogen in farts allow them to catch fire.

☞ Pure methane doesn't stink itself, but when it contains a little dimethyl sulfide and methanethiol... yuck!

☞ If you really want a fart to make people sit up and take notice, try foods such as cauliflower, eggs, and meat.

☞ On average, a person will produce a pint of fart gas every day. That's enough for about 14 farts.

Oh, Grow Up!

Our bodies are changing all the time, from the day we are born to the day we die.

TRUE BABY FACTS

☛ Babies have twice as many taste buds as adults.

☛ Baby boys grow faster than baby girls in the first seven months.

☛ In the first 238 days before birth, a baby's weight increases 5 million times.

☛ Inside the womb, a baby turns somersaults, and scratches itself with its fingernails.

☛ A baby's head is three-fourths of the size it will be as an adult—and one-fourth of its total body weight.

☛ A baby often knows when someone is speaking in a foreign language.

☛ Our eyes are always the same size from birth, but men's noses and ears never stop growing!

Diaper Delight

A newborn baby expels its own body weight in waste every 60 hours, which is why its diaper needs changing so much.

Growing and Puberty

➤➤ At the age of two, your brain contains twice as many connections between neurons—and consumes twice as much energy—as it will when you're an adult.

➤➤ Children grow faster in the springtime.

➤➤ You don't grow at the same speed all the time. Growth is quick in the first two years, slows down, then speeds up again in teenage years.

➤➤ At their fastest, boys can grow as much as 3.5 inches a year and girls by 3 inches a year.

TRUE OLDIE FACTS

☛ A baby girl in Australia can expect to live to 83 on average.

☛ In 1901, 1 percent of the world's people were over 60. Now, in Japan, 20 percent are.

☛ Proportionally, more people reach 100 on the Italian island of Sardinia than anywhere else.

☛ The British Queen has said happy birthday to 100,000 100-year-olds.

Oldest Person Ever

The oldest person whose age was proven was Jeanne Calment from France, who died in 1997 at the age of 122. Others may have lived longer, but their papers have been lost.

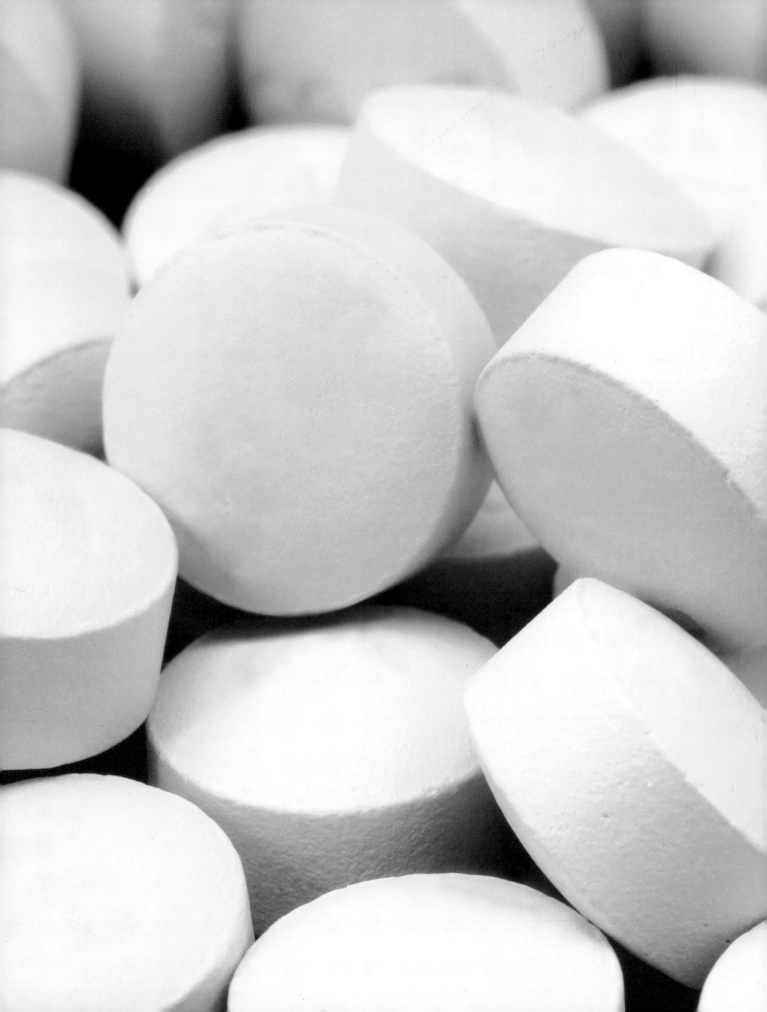

SICKNESS

AND HEALTH

There could be as many as 5 billion nasty viruses in one drop of blood. No wonder we get sick so often. In the past, people had some strange ideas about how to treat diseases, from drilling holes in skulls to pumping water up people's buttocks. Doctors know a lot more now, luckily for us!

Deadly Plague!

Quacks and Quirks!

Miraculous Medicines!

I Need a Dead Mouse, Doctor!

Doctors in the ancient world had some very strange ideas and came up with some truly wacky treatments.

Cupping

Two thousand years ago, doctors in ancient China applied heated cups to the skin to heal people. Cupping is still practiced today.

If you had an earache in ancient Mexico, Aztec doctors poured rubber in your ear. That way you wouldn't hear them when they asked, "Does it still hurt?"

TRUE HUMOR FACTS

☞ Many doctors in the ancient world believed the body was made entirely of four different kinds of substance called humors. But they were far from funny. They called these substances black bile, yellow bile, blood, and phlegm (mucus and snot).

☞ Until the 1600s, doctors thought illnesses were caused by an imbalance between the humors in the body. Their job was to restore that balance to make you better.

☞ Doctors thought if you were ill with a fever, it was because you had too much blood. So what did they do? They cut your veins to let it out, of course.

Having a hole drilled in your head was an ancient form of surgery called trepanning. Emergency trepanning is still carried out to relieve the pressure on injured brains.

Ancient Egypt

➤➤ You may think only pirates with boisterous parrots went around with replacement body parts made of wood. The ancient Egyptians replaced missing parts with wood, too, but only their toes.

➤➤ If you got a toothache in ancient Egypt, you might wonder if the treatment was worse than the pain. Egyptian doctors recommended holding a freshly killed mouse on your gums.

➤➤ If you were getting old and losing your sight, Egyptian doctors had a handy remedy—just rub your eyes with mashed tortoise brains and honey.

37

Quacks and Quirks

Really useless doctors are called quacks, but nobody is quite sure why.

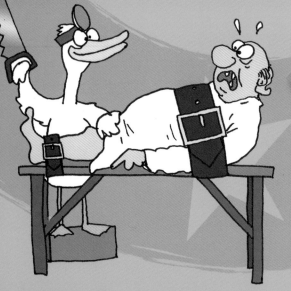

Quack, Quack

☞ Throughout the 1600s and 1700s, would-be doctors had to advertise their invariably useless cures by shouting at the top of their voices.

☞ Their squawking sounded so much like ducks that they came to be called quacks, some say. And every ducklike quack came with a bill...

Mercury

In the 1600s, mercury was used, wrongly, for treating diseases such as syphilis. The physician Paracelsus used it for treating rashes. Mercury was known as "quicksilver," or to some as "quacksalber." This may also be where the name quack comes from.

TRUE QUACK FACTS

☞ In the 1730s, Joanna Stephens came up with her own patent cure for bladder stones. Her "cure" was a mix of eggshells, soap, and honey, along with a few herbs. It was later thought the lime in the eggshells may actually have been beneficial, making urine more alkaline and dissolving the stones.

☞ In the 1800s, American doctor Samuel Thomson treated sick people by heating them up in steam baths. He gave them cayenne pepper and made them vomit by feeding them lobelia flowers.

Snake Oil

✎ In the American West of the 1800s, quacks tried to get rich by selling a magic remedy called "snake oil."

✎ Snake oil was an effective traditional Chinese remedy for painful joints, made from the fat of Chinese water snakes.

✎ When Chinese laborers brought it to North America in the early 1800s, quacks tried to make their own versions—without knowing any of the ingredients! The results were so useless that the term "snake oil" became a byword for useless medicine.

Dr. Kellogg

☛ Dr. John Harvey Kellogg (1852–1943), inventor of cornflakes, was a firm believer in healthy bowels. He thought the key to good health was to flush out the digestive system.

☛ Kellogg believed that 90 percent of all illnesses came from "putrefactive changes which recur in the undigested residues of flesh foods," which means that he thought disease was caused by parts of meat in poop.

☛ He treated his patients with a continual flow of water from both ends (in the mouth and up the buttocks). Kellogg could swish an amazing 16 gallons through somebody in just a few seconds.

In the 1820s, John Long rubbed ointment on the bodies of patients with tuberculosis (TB). If the ointment changed color, he thought the disease was being "extracted." He died of TB.

In the 1780s, British doctor James Graham promised to cure men of infertility by burying them up to their necks in warm earth. Amazingly, many men tried it.

Surgeons and Sawbones

In the days before anesthetics, surgery was a painful business.

Dr. Tagliacozzi

☛ If a man went around with his arm sewn to his head for a week in 16th-century Florence, you knew he'd been to Doctor Tagliacozzi.

☛ Tagliacozzi gave patients skin grafts for damage to their face using skin from their arm—but they had to spend a week holding the skin on while the graft took.

Amputation

➤➤ Having a limb amputated with no anesthetic put a patient through unimaginable agony.

➤➤ The quicker the operation, the less the patient suffered. So the best surgeons were those who could saw off a leg quickest.

➤➤ In the early 1800s, British surgeon Robert Liston could cut off a leg in just 28 seconds!

When the first amputation under anesthetic was done in London in 1847, the patient, butler Frederick Churchill, woke up after the surgeon had taken off his leg and said, "OK, doctor, when are you going to begin?"

Sawbones

☞ In the old days, surgeons were often called "sawbones," for obvious reasons.

☞ In Great Britain, surgeons are called Mr., Mrs. or Miss. and not Dr., because most surgeons were not originally qualified medics.

☞ Surgeons carried a large saw for cutting through leg bones, and a small one for cutting through arms.

Major Surgery

✏ In ancient India, men who were unfaithful to their wives were punished by having their nose sliced off. A clever surgeon named Susruta made a fortune by rebuilding their noses using a flap of skin taken from their forehead. These were the world's first skin grafts.

✏ In the 1800s, Frenchman Jean-Baptiste Denys gave the first blood transfusions. He used lambs' blood, his patients died, and Jean-Baptiste was arrested for murder.

✏ In the 4th century AD, saints Cosmas and Damian are said to have given Bishop Justinian a leg transplant to save him from blood poisoning.

The red and white striped barbers' pole came about because of the blood spilled by barber surgeons during operations. The white represented the bandage.

➤➤ Early surgery was often performed by barbers (hairdressers), who also wielded a nifty blade for shaving.

➤➤ The blades they used to shave their customers were called "cut-throat" razors. Ouch!

Deadly Plague

Centuries ago, Europe was hit by waves of terrifying diseases. They called these diseases "the plague."

TRUE PLAGUE FACTS

☞ One of the deadliest outbreaks of disease in history was the Black Death, an outbreak of the plague that swept across Europe between 1347 and 1351, killing over 25 million people—one person in every three. Whole towns were wiped out.

☞ Smaller outbreaks continued for centuries until people appeared to develop resistance.

☞ The Black Death may have been the disease bubonic plague, because one of the first symptoms was horrible lumps on the body called "buboes."

☞ Huge pus-filled buboes grew on victims' arms, necks, and groins. The buboes were caused by the swelling of lymph nodes, which are parts of the body that fight infections.

☞ In a rare, and particularly nasty, kind of plague called septicemic plague, people's skin turned deep purple or even black. They would die within hours of the symptoms appearing.

➤➤ Because they thought the plague was spread by bad air and touching victims, doctors visiting plague victims wore a long metal beak stuffed full of herbs. They also wore a thick leather gown and gloves.

➤➤ In fact, the plague could be spread by sneezing, which blew germs into the air.

Rats and Fleas

☞ Many people blamed rats for the plague, but it was actually carried by fleas.

☞ The plague germ, called *Yersinia pestis*, was passed from one rat to another by rat fleas. Large numbers of infected rats died. Only when the rats had died did the fleas turn to another source of food: humans.

☞ Fleas spread the plague germ when they bite their host.

☞ A hundred years ago in British India, you could get a job as a flea counter, counting how many fleas there were on a rat. A lot of fleas meant an outbreak of plague was imminent.

☞ Fleas have an internal thermometer that tells them when to leave a dying rat's body.

Great Plague of London

✎ A huge wave of bubonic plague hit London in the years 1665 and 1666. About 100,000 people are thought to have died or 20 percent of the city's entire population. This outbreak was called the "Great Plague."

✎ The disease was probably brought over on boats from Amsterdam. It spread quickly through the poor slums in the East End docks.

✎ The Great Fire of London on September 2–3, 1666 was a disaster that destroyed much of the city. But it also ended the Great Plague because it killed the city's rats and their fleas.

Dire Diseases

We live longer today than ever before, but killer diseases still stalk our world.

Smelly Smokers

☞ Smoking is not a disease but it is most certainly a killer.

☞ Three million people die of lung cancer every year, and 87 percent of those cancers are caused by smoking.

Two out of three people in rich countries die younger than they should because they eat too much fatty, salty, and sugary food.

➤➤ Every second of every day, someone new catches the lung disease tuberculosis (TB).

➤➤ TB kills more women than any other disease.

➤➤ One in three people around the world now has TB.

Malaria

✎ The disease malaria kills three people in Africa every minute. Children are the most common victims.

✎ Over 300 million people around the world have malaria each year.

✎ In Africa, someone is bitten every second by a mosquito infected by malaria.

✎ Drugs that can treat malaria are too expensive for most sufferers to buy.

Malaria means 'bad air' in Italian, and people once thought you caught it from the air in marshes. In fact, it is caused by a parasite that gets inside you when you are bitten by an anopheles mosquito.

Cholera

✎ The name of the disease comes from the Greek for diarrhea.

✎ You normally get cholera by drinking dirty water.

✎ Once in the guts, toxins in the cholera bacteria stop the gut from soaking up the digestive juices, causing watery diarrhea.

➤➤ One of today's most deadly diseases is **AIDS**, which is caused by the **HIV** virus. **HIV** stands for Human Immunodeficiency Virus.

➤➤ More than 22 million people worldwide have died of illnesses related to **AIDS**.

Measles Protection

☞ Measles is the world's most infectious, or easily spread, disease.

☞ If there is just one person infected with measles in a crowded room, everyone else in the room will catch it, unless they have been vaccinated.

Ghastly Germs

If you don't keep places clean, billions of tiny germs will soon make themselves at home.

Bacteria

☞ If you could get bacteria to line up single file, you could still fit in 10,000 across your fingernail.

☞ Although they live everywhere, bacteria are really at home in yucky places such as poop, or stuck to the feet and legs of animals such as cockroaches.

☞ Not all bacteria are bad. Many live harmlessly in your gut.

Germs

➤➤ Germs are the microscopic organisms that cause disease.

➤➤ There are three main kinds: bacteria, which cause diseases such as plague and tuberculosis; viruses, which cause diseases such as flu and yellow fever; and protozoa, which cause diseases such as malaria.

➤➤ The human body is a good home for many bacteria—just the right temperature and full of energy-giving sugar for them to eat.

➤➤ Bacteria often make a meal of flesh, but they have to get past the skin first.

46

☞ Viruses are tiny pieces of genetic material, which get inside living cells and start to reproduce.

☞ Scientists cannot agree whether viruses are living things or not. Viruses can reproduce, but only with the help of the cells they hijack.

☞ Viruses are so tiny you can see them only with a powerful electron microscope.

☞ A single drop of blood may contain as many as 5 billion viruses.

☞ Scientists are now looking at ways of treating diseases using viruses, which may in the future be used to kill harmful cancer cells.

Dirty Offices

Some scientists say the average office contains more than 400 times as many germs as a toilet. They estimate that there are 20,000 microbes in every square inch of the average desk. Other scientists say there are more on a car steering wheel.

➤➤ One sneeze can contain 6 million viruses. Millions of snot lumps containing viruses shoot out of your nose and mouth at high speeds.

➤➤ If your sneeze were a gust of wind, it would be strong enough to snap twigs.

Crummy Colds

The common cold is, as its name suggests, the most common illness of all. Not something to sneeze at.

Colds

✎ Over 250 different types of viruses can give you a cold.

✎ The most common cold viruses are called rhinoviruses, not because you catch them from rhinos, but because "rhino" is Greek for nose.

✎ To catch a cold, the virus has to get right up your nose. Some get up there as droplets in the air from coughs or sneezes. Others get on your hands, then you put your hands to your nose.

➤➤ The first sign of a cold is a sore throat. This is followed by a runny nose, coughing, and sneezing, which are your body's ways of fighting the infection.

➤➤ A high temperature in addition to these signs usually means you have the flu.

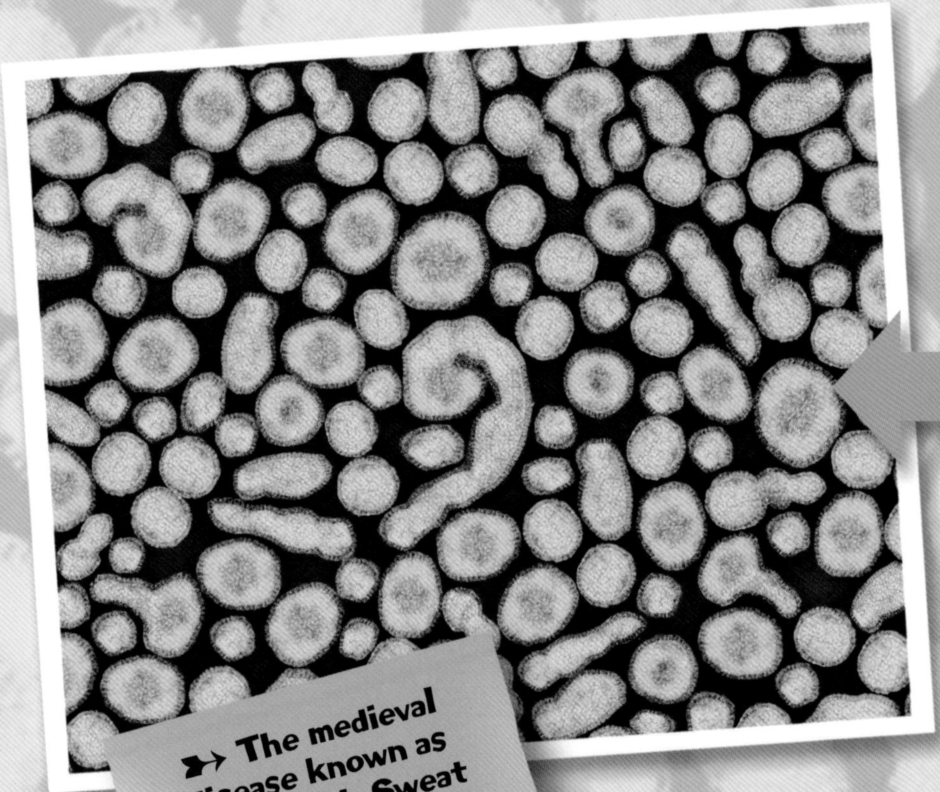

Flu Facts

☞ Flu, like colds, is caused by viruses such as those pictured here. Doctors can vaccinate you against many kinds of flu.

☞ Unfortunately, there are hundreds of different kinds of flu virus, and they are changing all the time.

☞ Each year a new range of viruses emerges to give people flu. That's why you can never be totally safe from it.

➤➤ The medieval disease known as the English Sweat was probably flu.

➤➤ The explorer Columbus's ships brought a flu epidemic to the Americas when they arrived in 1492.

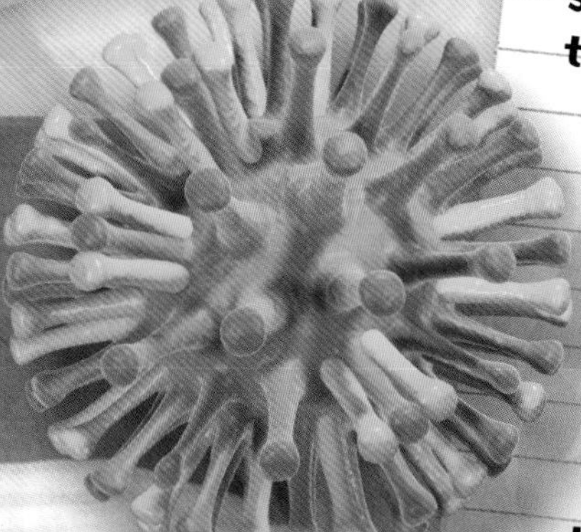

Flu Pandemics

➤➤ Every now and then very serious varieties of flu develop, sometimes mutating from viruses that normally only affect animals, such as these bird flu viruses.

➤➤ These new viruses sometimes cause a pandemic— a worldwide outbreak—that can kill hundreds of millions of people. Last century there were six pandemics of killer influenza: 1918, 1947, 1957, 1968, 1977, and 1989.

➤➤ The pandemic of 1918 came at the end of World War I. More people died during the pandemic than in the whole of the war.

49

Rotten Rashes

Your skin can come out in nasty spots and rashes, but they often look more serious than they are.

Anyone touching poison ivy, poison oak, or poison sumac plants is likely to develop blisters or a red itchy rash. The culprit is a nasty oil in the plants' sap called urushiol.

Ringworm

☞ Ringworm is not a worm at all, but a fungus that grows on the skin.

☞ It is called ringworm because the fungus causes ring-shaped rashes to develop.

☞ Warts are tiny swellings that grow mostly on your hands and feet and look like mini cauliflowers. They are mostly harmless and usually disappear after a few months.

☞ Warts are caused by a virus called the human papillomavirus (HPV).

☞ "Common warts" are rough little lumps that grow on your hands and your knees.

☞ "Plantar warts" are specky lumps, sometimes called verrucas, that grow on the balls and heels of your feet.

WART

Toads look like they're covered in nasty warts. But the growths on their skin aren't warts at all. It's not true that you can catch warts from touching a toad.

Awful Acne

☛ More than 90 percent of all teenagers get acne.

☛ Acne is the most common skin disease in the United States. Over 17 million people have it.

☛ Acne typically starts at around age 11 for girls and 13 for boys. Boys get it more, though, because of their changing levels of the male hormone testosterone.

☛ Acne develops as oil-oozing sebaceous glands in your skin get clogged with dead skin cells.

☛ On some parts of your face there are 1,200 glands in a square inch. Each gland could grow into an acne spot.

Painful Boils

✎ Boils are more serious than acne. They erupt when *Staphylococcus* bacteria get in through a break in the skin to infect a blocked gland or hair root.

✎ When bacteria get in, skin cells try to fight them off. This makes the spot red and itchy.

✎ The blood brings in a mass of white blood cells to attack the germs, and the battle leaves a debris of pus, which swells the spot into a painful boil.

✎ Boils rising on a tight area of the skin, such as the back of the neck, may band together into a horrible, multiheaded monster called a carbuncle.

Miraculous Medicines

Most of the effective medicines we use today were inspired by nature.

TRUE ANTIBIOTIC FACTS

☞ Antibiotics are amazing drugs that attack harmful bacteria without harming the patient. Antibiotics have saved countless lives.

☞ Many antibiotics are based on substances taken from mold.

☞ Almost one-fourth of a billion doses of antibiotic are taken by people each year. Many more are given to farm animals.

☞ If antibiotics are used too much, bacteria can develop resistance to them, making them useless.

Alexander Fleming discovered the first widely used antibiotic, penicillin, by accident in 1928, when he saw how bacteria died in a dish containing penicillin mold.

Leeches

☞ In the 1800s, doctors used worms called leeches to suck patients' blood.

☞ Now surgeons are using leeches again when performing a skin graft. A chemical in the leeches' saliva stops the blood from clotting.

Steroids

➤➤ Steroids are natural substances the body uses to keep itself in balance. Steroid drugs mimic them to produce the same effects.

➤➤ Corticosteroids are amazing drugs that stop parts of the body from becoming inflamed. They are used to treat illnesses, such as arthritis, which causes joints to swell.

➤➤ Anabolic steroids promote cell growth, and are used by cheating sportspeople to boost performance.

Aspirin

☛ Aspirin is based on a natural chemical found in willow bark.

☛ The ancient Greeks were using willow bark to ease pain and fever 2,500 years ago.

☛ Aspirin is made from petroleum nowadays.

☛ Over 100 billion aspirin pills are taken around the world each year!

Delighted Dairymaids

In the 1700s, dairymaids were known for their smooth skin, unblemished by smallpox. It may be that their cows gave them a mild version of the disease, called cowpox. So their bodies developed the antibodies to fight off smallpox.

➤➤ When dried, figs are very high in fiber. Eating extra fiber may help your body ward off diseases by helping food pass through your guts more smoothly.

➤➤ However, too much fiber in your food makes you fart, and green figs can give you diarrhea.

53

Smooth Operators

Surgery has come a very long way since the invention of anesthesia 150 years ago. Soon it could all be done by robots!

Eye Surgery

☞ Poor eyesight can often be corrected by shaving the lens of the eye to the right shape using laser beams.

☞ Very occasionally, laser eye surgey goes wrong and a cornea transplant is needed. Cornea-harvesting robots cut a replacement perfectly from a deceased donor.

➤➤ If you had an operation in the early 1800s, the surgeon often killed you by passing on an infection with his dirty hands.

➤➤ Today, protecting patients against germs is a top priority.

Anesthetists often inject patients with a substance called curare. This is very similar to the poison used on poison arrows by the people who live in the Amazon rainforest.

☛ The da Vinci system is a tiny robot that can enter the body through a small slit to perform operations with the minimum upset. The surgeon controls the robot using a computer screen.

☛ By using technology such as fiber optics and satellite links, surgeons can now conduct "remote surgery" from thousands of miles away.

☛ In May 2006 in Milan, Italy, a robot carried out a simple heart operation on its own, while surgeon Carlo Pappone went off to a conference.

Laparoscopy

✎ To perform a laparoscopy, surgeons insert their instruments into the body through just a tiny slit in the outside. It's a bit like making your bed with a wire pushed through the keyhole of your door.

✎ The laparoscope is the video camera device that shows the surgeon what's going on inside the patient's body.

✎ Doctors in Great Britain call laparoscopy keyhole surgery.

Super Scans

☛ In MRI (Magnetic Resonance Imaging) scans, the patient slides through a ring of magnets that help create a picture of inside their body.

☛ Using PET (Positron Emission Tomography) scans, scientists can track blood through a live brain.

☛ Surface-enhanced Raman spectroscopy allows scientists to instantly scan a body for viruses such as HIV.

Good as New

We can now replace many body parts, and could soon be taking them from animals.

Pigs could provide hearts for transplant if they can be genetically engineered so that human bodies don't reject the heart. This would meet the ever-growing demand.

TRUE HEART FACTS

☛ In the 1890s, a French surgeon, Alexis Carrel, transplanted a dog's heart, putting the heart in the poor animal's neck.

☛ The first surgeon to successfully transplant a human heart was Christiaan Barnard in Cape Town on December 3, 1967. The patient survived for 18 days.

☛ In 1981, Dr. Bruce Reitz of Stanford University of Medicine performed the first transplant of both the heart and lungs.

☛ In 2005, over 80,000 people around the world had heart transplants.

☛ In the United States, there are six heart transplants every day.

☛ Five out of six heart transplant patients survive at least a year, and usually much longer.

☛ In 1969, a patient was given an electric heart to keep him alive while a donor heart was found. Temporary hearts like these are called bridges. Many patients now live for months with electric hearts.

Terrific Transplants

☛ A transplanted organ may be "rejected" because the body's immune system thinks it is foreign and attacks it.

☛ After a transplant, patients are given drugs, such as cyclosporin and steroids, to stop the organ being rejected.

☛ Doctors can now transplant all the major body organs—heart, lung, kidney, pancreas, and liver.

☛ In 2005, a woman in France who had been attacked by a dog was given a partial face transplant (right).

☛ In 1998, French surgeon Jean-Michel Dubernard transplanted a new hand onto a man who had lost his in a circular saw accident.

☛ Scientists are now experimenting with transplanting animal parts into humans.

The Future...

✎ Professor Robert White, of Cleveland, Ohio, transplants the heads of monkeys. He thinks he could one day transplant human heads.

✎ In a technique called tissue engineering, new organs for transplant could soon be grown on frameworks in a laboratory.

Bionic Bodies

An alternative to taking parts from other bodies is to make completely artificial ones.

The battery-powered "Utah Arm" moves in response to the owner's thoughts. Its movements are controlled by a computer that responds to muscle twitches in the stump of the owner's real arm.

Artificial Organs

➤ While hiding from the Nazis in World War II, Dutch doctor Willem Kolff made the first artificial organ—an artificial kidney—from old sausage skins, orange juice cans, and an old washing machine.

➤ Artificial kidney machines are very big and patients must sit still while connected to them. Scientists are now developing an artificial kidney that will fit inside the body.

➤ Doctors are testing artificial retinas for blind people. The patient has glasses fitted with a tiny camera that feeds its picture to the new retina, which sends the picture to the brain.

Prosthetics

☞ Artificial replacement body parts are called prosthetics.

☞ For centuries, prosthetic limbs were just things like wooden legs. Now, using special materials like titanium and Kevlar, scientists make sophisticated prosthetic limbs.

☞ In 2004, a dolphin in Japan was given a prosthetic fin.

☞ Bionic is a term used for complete replacement body part systems. In 2002, American Jesse Sullivan was fitted with the first bionic arm, which he could control using his chest muscles.

☞ The bionic arm detects movements in the chest muscle, which is connected to nerves that once went to the real arm.

Rudy García-Tolson

☛ Even though he had both legs amputated above the knee when he was five, teenage athlete Rudy Garcia-Tolson plays football and swims. He even runs on the track.

☛ Rudy Garcia-Tolson wears different prosthetic legs for sports and for everyday.

☛ Rudy's legs don't look real, but they are made of carbon fiber that bends, springs, and absorbs shock just like real legs.

☛ He won a swimming gold medal at the 2004 Paralympics at the age of just 15.

The Future

✎ Scientists believe that one day they will be able to make replacements for just about every body part except the brain.

✎ In the future, more and more bionic parts will be linked directly into the brain, so that they can be controlled by thought alone.

✎ Some new body parts will be made of special materials that are even tougher than natural ones.

✎ Tiny computers may be implanted into your brain to help you think.

THEY DID WHAT?!

Here are some of the bravest and cleverest people in history, along with a pick of the weird and the wrong. Meet heroes like Robert Bartlett, who walked 600 miles across the Arctic to find help. But stay away from the likes of Spanish prince Don Carlos, who once forced a shoemaker to eat a pair of boots!

Real Superheroes!

Smooth Criminals!

Brainy or What?!

Real Superheroes

There are many ways to be a hero. To some, great conquerors are heroes. Others prefer brave explorers or champions of the poor.

Classical Conquerors

☞ The great ancient Roman leader, Julius Caesar, left, (100–44 B.C.) who conquered many lands, was said to have been captured by pirates when on his way to school as a boy, but charmed his way to freedom.

☞ Ancient Greek leader, Alexander the Great (356–323 B.C.), forged an empire covering 1½ million square miles over three continents.

☞ Alexander was tough, but when he was attacking Petra in present-day Jordan, he had ice brought from the mountains each day so he'd always have cool drinks.

TRUE REBEL FACTS

☞ When Scottish rebel William Wallace (1270–1305) was executed, his right arm was taken to Newcastle and his left to Berwick, his right leg to Perth and his left to Aberdeen. His head was stuck on a spike in London.

☞ After a heroic campaign to unify Italy, Garibaldi (1807–1882) went out for a pizza to celebrate... really!

☞ Haitian independence leader Toussaint l'Ouverture, right, (1744–1803) was known as "The Opener" because even when he was trapped, he would find an opening.

George Washington (1732–1799) was the first president of the United States and the only one ever to be elected unanimously. But he had to borrow the money to go to his own inauguration!

Simón Bolívar

The country Bolivia is named after the hero Simón Bolívar (1783–1830). Bolívar led the forces that ousted the Spanish from much of South America.

Intrepid Explorers

☛ When Christopher Columbus landed in Cuba in 1492, he was convinced he'd gone all the way round the world and reached India. That's where the West Indies got their name.

☛ Ferdinand Magellan (1480–1521), the explorer who captained the first ship to sail around the world, did not make it home. He was killed by the people he found in the Philippines. Only 18 out of his 270 crew made it all the way.

☛ The famous explorer of Australia, Captain James Cook, was the first European to see surfing. People were surfing when his stopped in Tahiti in 1769.

☛ When exploration ship *Karluk* was crushed in Arctic ice in 1913, Robert Bartlett walked 600 miles across the Arctic and Siberian ice to get help.

Magnificent Medic

✎ Che Guevara (1928–1967) was an Argentinian doctor who championed the poor and fought in the guerrilla war in Cuba that brought Fidel Castro to power.

✎ Many people who like what he stood for wear Che T-shirts, usually showing the famous photo of him by Alberto Korda.

Big-hearted People

Throughout history, there have been people who devoted their lives to improving conditions for others.

Marvelous Men

➤➤ **Mahatma Gandhi (1869–1948)** wanted to free India from British rule by peaceful means. To protest against the British salt tax, he led thousands on a 250-mile walk to get salt.

➤➤ **Tenzin Gyatso** is the 14th Dalai Lama, leader of Tibetan Buddhists. Exiled to India, he campaigns peacefully to free Tibet from Chinese rule.

➤➤ German doctor **Albert Schweitzer (1875–1965)** developed a philosophy based on respecting all life. He devoted his life to helping the sick in Africa and campaigned against nuclear weapons.

Front-line First Aid

Nurses Mary Seacole and Florence Nightingale braved the horrors of the Crimean War (1853–56) and revolutionized the treatment of injured soldiers.

Francis of Assisi

➤➤ St. Francis of Assisi (1182–1226) saw all animals as brothers and sisters. Stories tell how he preached to birds.

➤➤ Francis is said to have made peace with a dangerous wolf. Part of the deal was that the town dogs would leave the wolf alone.

TRUE FEMALE FACTS

☛ Shepherd girl Joan of Arc, right, (1412–31) led the French army at just 16 years of age to rescue Orléans from the English. The English burned her alive for the "crime" of wearing men's clothes!

☛ Alabama-born Helen Keller (1880–1968) was both deaf and blind, but she worked tirelessly for the disabled and the poor, saying: "I have visited sweatshops, factories, and slums. If I could not see it, I could smell it."

☛ Albanian nun Mother Theresa (1910–1997) dedicated her life to the poor of India, founding orphanages and leper colonies.

In 1994, Nelson Mandela became the first president elected by all South Africans. He had been imprisoned by the country's white-only apartheid regime for 27 years.

Fight for Your Rights

☛ Black woman Rosa Parks (1913–2005) changed U.S. history in 1955 when she wouldn't give up her seat on a bus to a white man.

☛ U.S. civil-rights leader Martin Luther King (1929–1968) was murdered because he fought for the rights of black people.

☛ W.E.B. Du Bois (1868–1963) campaigned for the rights of all those descended from Africans.

Dreadful Despots

Some bloodthirsty, ruthless, and simply dreadful rulers.

Ancient Tyrants

➤➤ It is said that Herod the Great (74–4 B.C.) killed all Bethlehem's boys in an attempt to kill Jesus.

➤➤ Ibrahim the Mad (1615–1648) was once so angry he had all his 280 wives thrown in sacks into the river.

➤➤ Mahomet IV of Turkey (1648–1687) had a clerk to keep a diary of his reign. After one dull day, the diary was blank. So Mahomet speared the clerk, saying: "Now you have something to write about."

➤➤ There were no flies on King Pepi II of Egypt (2284–2184 B.C.). He had slaves stand near him covered in honey to draw the flies away.

Elagabalus

☛ Roman emperor Elagabalus (203–222) held wild parties where guests would dine on live parrots.

☛ Dinner guests would sometimes be suffocated when he filled the room with flower petals.

☛ His enemies would have their skin stripped before a stinging dip in salty water.

☛ He displayed the body parts of people he had killed in golden bowls.

Commodus

☛ Roman emperor Commodus (A.D. 161–192) liked to fight gladiators, and charged the city of Rome so much money for his appearances in the arena that it put a strain on the economy.

☛ Eventually, Romans could not stand him any longer, and he was strangled in his bath by the wrestler Narcissus.

Ivan the Terrible

>> The first emperor of Russia, Ivan (1530–1584) was called Terrible because he was scary, not because he was just no good.

>> In 1570, Ivan was so sure that the people of the city of Novgorod were traitors, he had the city burned down and all its people killed.

>> He accidentally killed his own son in the heat of an argument.

>> Ivan blinded the men who built St. Basil's Cathedral for him so that they couldn't build anything better.

Juan Rosas

☞ Juan Manuel de Rosas (1793–1877) unified Argentina, but he was a ruthless tyrant who once ordered the execution of a very pregnant woman.

☞ Rosas insisted people show their support for him by wearing a red ribbon at all times.

Robespierre

>> Maximilien Robespierre (1758–1794) was a leader of the French Revolution so dedicated to the cause he became known as "The Incorruptible."

>> He was very absent-minded, once spilling soup onto a table after failing to notice that he didn't have a bowl.

>> Robespierre started the mass executions of his enemies known as the Reign of Terror, in which thousands of people had their heads sliced off by the guillotine. In the end, he was himself guillotined.

Crackpot Royalty

When everybody has to do as you say, you can get away with some very strange behavior.

Regal Excess

☞ English King Henry VIII (1491–1547) had a servant whose job was to wipe his bottom when he pooped.

☞ The son of Spanish king Phillip II, Don Carlos (1545–1568) once forced a shoemaker to eat a pair of boots that were badly made.

☞ Ludwig II of Bavaria (1845–1886) was obsessed with swans, and surrounded himself with pictures and statues of them.

Riotous Romans

➤➤ Nero, left, (A.D. 37–68), is said to have killed his mother Agrippina so that she could not prevent him divorcing his wife and marrying his friend's wife.

➤➤ When his sister died, Caligula (A.D. 12–41) was so upset, he banned people from laughing on pain of death. He also banned them from taking a bath. What a stinker!

➤➤ To show the Senate what he thought of them, Caligula made his horse Incitatus a senator.

➤➤ The food at Caligula's dinner parties must have been bad. To keep guests happy, he would behead a criminal or two between courses.

➤➤ Gluttonous Emperor Vitellius (A.D. 15–69) ate four banquets a day, and once ate a pie the size of a room. He died by choking on the beak of a crow he was eating for dinner.

TRUE EMPEROR FACTS

☞ Holy Roman Emperor Rudolf II (1552–1612), whose crown is shown here, collected dwarfs and giants.

☞ Napoleon Bonaparte of France (1769–1821) surrounded himself with very tall guards, and this may be why people thought he was short. In fact, he was probably around 5 feet 6 inches tall, which was about average for his time.

The Madness of George III

British King George III (1738–1820) had a mental illness. For a time, he ended every sentence with the word "peacock." He also sometimes spoke for many hours without pause, and claimed to talk to angels.

Jean-Bedél Bokassa (1921–1996) modeled himself on Napoleon and declared himself "Emperor of Central Africa." He bankrupted his country with his inauguration ceremony.

➤➤ Every royal needs a hobby. Britain's Prince Charles is a keen gardener and likes to talk to his plants.

➤➤ Henry VIII loved tennis, but he was so fat he needed a servant to throw the ball up for him.

Crafty Creatives

Artists, writers, and composers can be an odd bunch. They aren't always the kind of people you'd like to introduce to your parents.

Colorful Composers

☞ Beethoven (left) dipped his head in water before he composed.

☞ Beethoven was said to wash all the time, but then wear dirty clothes.

☞ Just before he became fatally ill, Mozart was asked to write a requiem—a huge funeral piece—by a mysterious stranger. He came to believe the funeral was his own.

☞ Frederic Chopin wore a beard on only one side of his face.

☞ Sir William Gilbert doted on a bee called Buzfuz.

☞ Penniless composer Eric Satie said, "People in a bar will always buy you a drink. But they'd never dream of giving you a sandwich."

Oddball Writers

➤➤ The ancient Roman poet Virgil was so upset when his pet fly died that he gave it a funeral.

➤➤ The poet Tennyson liked to do an impression of a man sitting on the toilet.

➤➤ The poet Swinburne once slid down the banister naked at a dinner party.

➤➤ The writer Oscar Wilde took a dead lobster for walks on a piece of string.

Shakespeare the Wordsmith

☞ When Shakespeare didn't have the right word, he invented it. In all, he invented 1,700 new words.

☞ He invented these words: assassination, bump, lonely, madcap, watchdog, gossip, blushing, barefaced, hobnob, bedroom, torture, undress, moonbeam, and zany.

TRUE ARTIST FACTS

☞ After a fight with fellow artist Gauguin, Van Gogh cut off his own ear in remorse.

☞ When he became poor, Whistler would paint pictures on the floor of all the furniture carted off by the bailiffs.

☞ L.S. Lowry always wore shabby suits. Asked what he did with his old clothes, he said, "I wear them."

☞ Salvador Dali (right) had an intense fear of grasshoppers.

☞ When he was young, Dali had a pet bat. It was eaten by ants, so Dali developed an obsession with ants.

Smooth Criminals

Thieves are sometimes glamorous figures, defying the law and refusing to live conventional lives.

Outside the Law

➳ The legend of outlaw Robin Hood tells how he lived in Sherwood Forest, England, in the 1300s, and robbed from the rich to give to the poor.

➳ Most scholars think Robin Hood never existed, but recent archaeological discoveries suggest he might actually have been real.

➳ Ned Kelly was the most famous of the bushrangers, outlaws who roamed the Australian bush in the 1800s, defying the law and looking for adventure.

Highway Robbery

☛ Highwaymen attacked stagecoaches in England in the 1700s and 1800s, crying, so legend has it, "Stand and deliver."

☛ "Swift Nick" Nevison earned his name by riding 200 miles from Kent to York in just 14 hours to provide himself with an alibi for a robbery.

☛ In Harrison Ainsworth's novel *Rookwood*, Swift Nick's ride was attributed to another highwayman, Dick Turpin, on his horse Black Bess. In reality, Turpin was an incompetent highwayman.

☛ Once when Claude du Vall stopped a coach, he played music with a lady passenger—then insisted her husband give him about $800 for the entertainment.

Gun-toting Gangsters

☞ Bonnie Parker and Clyde Barrow were lovers-turned-gangsters in the US in the 1930s. They became celebrities for their daring robberies.

☞ Notorious Chicago gangster Al Capone earned the nickname Scarface at the age of 15, when his face was slashed by someone he insulted in the bar where he worked.

Wild West

➤➤ Butch Cassidy and the Sundance Kid were members of a gang called The Wild Bunch, who held up banks and trains in the early 1900s.

➤➤ Harry Longabaugh was called Sundance Kid because he was caught stealing a horse in Sundance, Wyoming.

➤➤ Gunfighter Wild Bill Hickock was killed holding a poker hand of aces and eights. This hand is now known as the Dead Man's Hand.

Ronnie Biggs was convicted for his part in the 1963 $4.8 million Great Train Robbery, in Great Britain. He escaped from jail and fled to Brazil, where this photo was taken.

TRUE THEFT FACTS

☞ Brilliant 19th-century bank robber Adam Worth stole a valuable painting by Gainsborough, then he had an attack of conscience and gave it back—25 years later.

☞ American bank robber Pretty Boy Floyd was famous for miraculously shooting his way out of police ambushes without a scratch.

☞ John Dillinger earned the nickname Jackrabbit for his way of leaping gracefully over the counter, or nimbly away from the police, when he robbed banks in the 1930s.

Stinking Rich

If you're rich enough, you can buy yourself almost anything.

Howard Hughes

☛ Son of an oil tycoon, Howard Hughes (1905–1976) used his huge wealth to make himself a film director and flying ace.

☛ When his wife didn't want to keep a stray cat she had found, Hughes sent it to a hotel where it had its own room. His secretary wrote the cat letters every month.

☛ When he ate cake, Hughes insisted it be cut into perfect squares, measured with a ruler.

Louis the Sun King

Known as the Sun King because he thought he was a god, Louis XIV (1638–1715) built the Palace of Versailles at a cost of one-fourth of all France's annual income.

Crazy Castle

☛ Hearst Castle was the house built by newspaper billionaire William Randolph Hearst in the 1920s in San Simeon, California, at a cost of about $1 billion in today's money.

☛ It is one of the most extravagant houses in the United States, with 56 bedrooms, 61 baths, and 41 fireplaces.

TRUE LUXURY FACTS

☞ This Bentley (right) is very expensive at $350,000, but it's a snip compared to the $610,000 SSC Ultimate Aero and the $1 million Bugatti Veyron.

☞ A flat now being built near London's Hyde Park will cost over $150 million!

☞ If you want to hire the Greek super-yacht *Annaliese* for just one day it will set you back over $110,000.

☞ One meal at Masa's in New York will cost you $550.

☞ A night at the Mansion at the MGM Grand Hotel, Las Vegas is a mere $4,600.

➵ Indonesian Kopi Luwak coffee costs $50 a cup. To make it, weasel-like palm civets eat the beans, and poop them out.

➵ London shop Selfridges' McDonald sandwich, filled with special Japanese Wagyu beef, is only for the upper crust at $150.

Flashy Accessories

➵ $550,000 for a Constantin Vacheron watch is a bargain compared to the diamond encrusted Chopard watch that recently sold for $24 million!

➵ The platinum-bodied, diamond-encrusted Diamond Crypto Smart phone made in Moscow costs a mere 1,300,000. Calls are extra.

➵ Stuart Weitzman shoes on sale at Harrods in London are woven from platinum thread and set with rubies and opals. They cost $1,300,000 – that's $650,000 per foot.

➵ The German firm Trekstor have created a diamond-crusted, gold-plated MP3 player—yours for just $26,000!

Gruesome Killers

The most notorious murderers in history killed for money, for love, or simply for the fun of it.

TRUE VLAD FACTS

☞ You may think that Count Dracula is just a horror story, but he was inspired by Vlad Dracul, who lived here at Bran Castle.

☞ He was also known as Vlad the Impaler because he had wooden spikes rammed through over 20,000 people.

☞ Vlad once invited the ill and poor to a banquet. He locked them in and burned down the banqueting hall.

☞ When Turkish guests kept their hats on during dinner, as was their custom, Vlad had the hats nailed to their heads.

Jack the Ripper

➤➤ Jack the Ripper is the name given to the unknown killer of at least five young women in London's East End in the 1880s.

➤➤ Police received a boasting letter from a man claiming to be the killer. It was signed Jack the Ripper and the name stuck.

➤➤ There were many theories about the Ripper's identity. Writer Arthur Conan Doyle thought it was a woman, and called her "Jill the Ripper."

Resurrection Men

✎ Today, people donate their bodies to medical science. But in the 1700s and 1800s, doctors and scientists had to buy bodies from "body snatchers," who stole corpses from graveyards.

✎ Body snatchers were sometimes jokingly referred to as Resurrection Men.

✎ Sometimes they were too impatient to wait until the victim was dead, so they sped him on his way by murdering him.

✎ The most notorious of these murderers were William Burke and William Hare, who prided themselves on their supply of freshly killed bodies to Edinburgh's medical students in the 1820s.

Serial Killers

☞ Since Jack the Ripper's time, there have been at least 100 serial killers around the world.

☞ One of the worst serial killers of modern times was Russian Andrei Chikatilo, who killed 53 women and children between 1978 and 1990.

☞ Pedro Alonso Lopez killed 300 young girls in Colombia, Peru, and Ecuador in the 1970s and 1980s.

☞ British doctor Harold Shipman was convicted in 2000 of killing 236 patients.

By the time his wife's body was found in London in 1910, murderer Dr. Crippen was sailing for Canada. He was the first criminal ever to be arrested with the aid of radio.

True Survivors

People have survived at times when it seemed they were certain to die. It shows what we can do when we have to.

Survival at Sea

➤➤ In 1799, the crew of the ship *Bounty* mutinied and set its captain, William Bligh, and some others adrift in the Pacific in a small boat. Bligh led his men to safety across 3,400 miles of ocean.

➤➤ Ernest Shackleton's ship *Endurance* was crushed in the ice of the Antarctic in 1915. Shackleton and a few of his men sailed 750 miles in a small open boat in the world's roughest seas to get rescue—and succeeded.

➤➤ After their yacht was capsized by a whale in the Pacific in 1973, Maurice and Maralyn Bailey survived 117 days in a rubber dinghy, eating fish caught using a safety pin as a hook.

Mission Accomplished

On a secret mission in World War II, American pilot Eddie Rickenbacker had to ditch his plane in the sea. He was found alive after a month in an open boat. His mission was successful.

The Real Crusoe

☞ Daniel Defoe's story of Robinson Crusoe was inspired by Alexander Selkirk, a British sailor who survived for four years on a desert island until he was picked up in 1709.

☞ At first, he lived in a cave on the beach, but he was scared inland by sea lions. In the interior of the island, he ate goats and tamed wild cats to keep him company.

☞ Selkirk had to hide from the first two ships he saw. They were Spanish, Britain's enemies, and would have killed him.

When an explosion disabled *Apollo 13* as it neared the Moon in 1970, its crew overcame crippling cold, low oxygen, and lack of water to return to Earth safely.

Survival in the Andes

When a plane crashed high up in the Andes in 1972, 16 passengers survived on the mountain for 72 days by eating the flesh of crash victims. Two men, Roberto Canessa and Nando Parrado, walked ten days to get help.

Extreme Measures

✏ Joan Murray was skydiving in North Carolina in 1999, when her parachute failed to open. She landed on a mound of stinging fire ants, whose stings kept her heart beating and saved her life.

✏ When a huge boulder fell on Aron Ralston in Utah in 2003, trapping his hand, he had to cut off his hand with a penknife to escape.

✏ When Vietnamese fisherman Bui Duc Phuc was out fishing in May 2004, a strong current swept him far out to sea. He survived for 14 days by drinking his own pee.

TRUE SURVIVAL FACTS

☞ In 1985, when Joe Simpson fell in the Andes, breaking his leg, climbing partner Simon Yates tried to lower him on ropes. But a slip left him dangling over a cliff, so Yates cut the rope to stop them both from falling. Certain Simpson was dead, Yates went on alone. But Simpson survived the long fall into a glacier, and crawled back to base over three days.

☞ When Mauro Prosperi ran a marathon in the Sahara in 1994, he got lost in a windstorm and had to struggle 125 miles to safety, surviving for ten days without water by drinking the blood of two small bats.

➤➤ In May 2005, 88 migrants were shipwrecked in the Pacific. They really were rescued when coast guards found a message in a bottle.

Brainy or What?

Almost superhuman or half crazy, the greatest minds in history have had a different way of looking at the world.

Geeky Greeks

➤➤ The ancient Greek thinker Plato was a perfectionist. He rewrote the opening to his book *The Republic* at least 50 times.

➤➤ Plato wrote about the mighty nation of Atlantis, now lost in the sea. People have looked for the "real" Atlantis ever since, but he made it up.

➤➤ The thinker Diogenes lived in a bath and behaved like a dog in public, peeing on those he disliked and pooping in the theater.

➤➤ When the king Alexander the Great asked frail old Diogenes if he could do anything for him, Diogenes replied, "Stand out of my sunshine."

➤➤ After hearing Allegri's complex work Miserere just once, Mozart (above) remembered every note.

➤➤ Pianist Franz Liszt played 21 three-hour concerts entirely from memory, without repeating a piece.

TRUE GENIUS FACTS

☞ Leonardo da Vinci was so far ahead of his time, he created designs for a parachute, a helicopter, a tank, a machine gun, and aircraft landing gear—in the 1600s! He was left-handed and wrote backwards so he wouldn't smudge the ink.

☞ The great scientist Isaac Newton, better known for his maths and physics, also invented the cat flap. He had two: one for his cat and one for her kittens.

☞ American scientist and statesman Benjamin Franklin invented the rocking chair, flippers, and bifocal lenses.

Leave on the Light

Inventor Thomas Edison was afraid of the dark, which explains why he helped invent the light bulb. Among many other things, Edison also invented a machine to record sound, the phonograph (right).

☛ Marie Curie (left) discovered radiation, and she was the first person to win two Nobel prizes.

☛ Albert Einstein's parietal lobe, the part of the brain that does math and logic, was 15 percent bigger than average.

☛ Despite being virtually paralyzed by a nerve disease since his mid-20s, physicist Stephen Hawking wrote the best-selling science book ever, *The Brief History of Time*, explaining complex theories of the universe.

TRUE BOFFIN FACTS

Young Master

Becoming a grand master is the highest level a chess player can reach, and there have been only 900 since the first in 1914. The youngest ever was Ukrainian Sergey Karjakin, who became one in 2002 at the age of just 12 years 7 months.

Blunders

Stupid mistakes have lost people money, a football match, and even their lives.

Famous Last Words

☛ "They couldn't hit an elephant at this distance" were the last words of Colonel John Sedgwick, Union Commander in the American Civil War, moments before he was killed by sniper fire at Spotsylvania in 1864.

☛ "I just can't get to sleep" were the last words of Peter Pan author J.M. Barrie, shortly before he went to sleep permanently.

In the Battle of Little Bighorn in 1876, Lt Col George Custer's troop was annihilated by Lakota-Cheyenne braves. Custer's famed "Last Stand" was a massive tactical blunder.

TRUE BUSINESS FACTS

☛ In 1991, when Gerald Ratner of jewelry company Ratners' in the UK was asked how he could sell jewelry at such low prices, he admitted that his goods were poorly made. Sales plummeted and he had to leave the company 18 months later.

☛ In 1991, vacuum cleaner company Hoover offered free flights to Europe and New York to anyone who bought $185 of Hoover products. Many of the flights cost more than $185 and the offer cost Hoover $88 million.

☛ In 2005, a trader in Japanese financial firm Mizuho made a disastrous typing error and sold 610,000 shares for 1 yen, not 1 share for 610,000 yen. It cost Mizuho about a billion dollars.

Snatching Defeat from Victory

Just yards from the winning post at the 1956 UK Grand National, race leader Devon Loch jumped an imaginary fence then collapsed in a heap.

Sport

➤➤ In the 1993 Super Bowl, Dallas Cowboys' Leon Lett was already celebrating a touchdown when he was tackled by an opponent. He failed to score.

➤➤ In 2006, Bury soccer player Chris Brass broke his nose attempting an overhead clearance—and scored an own goal.

Bushisms

➤➤ George W. Bush (below), famous for his verbal gaffes and sounding confused, once said, "Those who enter the country illegally violate the law." Hmmm, think about it, Mr. Bush.

➤➤ Speaking of how people of the future would judge him, he said, "You can never know your history until after you're gone."

Silly Scientists

☞ In the 1920s, bones found in England, known as Piltdown Man, were hailed as the missing link between apes and men. In 1953, the skeleton was revealed as a hoax made from a human skull and an orang-utan jaw.

☞ For 70 years, spinach was thought to contain 10 times as much iron as it does, due to a math error by Dr. E. van Wolf.

☞ In 2006, scientists thought that the cattle brain disease BSE had jumped into sheep. It turned out they had been testing the brains of cows, not sheep, by mistake.

83

Outstanding Athletes

By pushing themselves to extremes, sport stars have achieved amazing feats.

In ancient Greece, athletes competed in the nude. The word gymnast comes from the Greek for "naked."

Athlete Jesse Owens won four track and field gold medals in the 1936 Berlin Olympics, a feat that was not matched until Carl Lewis did the same in 1984.

Babe Ruth

☛ Baseball striker George "Babe" Ruth was also known as The Bambino, the Sultan of Swat, and the Colossus of Clout.

☛ He got the nickname "Babe" as he was called "Jack's newest babe" when he was first signed to manager Jack Dunne's Baltimore Orioles in 1914.

☛ Ruth hit a record 60 home runs in a season in 1927. The record stood for 34 years.

☛ In 1998, *The Sporting News* voted Babe Ruth the greatest ever baseball player.

☞ Stella Walsh won the women's 100-metre gold for the U.S.A. in the 1932 Olympics. But when she died in 1981, the autopsy revealed she was a man.

☞ Jackee Joyner-Kersee was the greatest female athlete of the 20th century. She won six Olympic gold medals and dominated the grueling seven-event heptathlon.

☞ Mildred "Babe" Didrikson was a brilliant all-round sportswoman, beginning with basketball, winning two running golds at the 1932 Olympics, then excelling at tennis, baseball, golf, and even billiards.

☞ The 14-year-old Romanian gymnast Nadia Comaneci scored a perfect 10 at the 1976 Montreal Olympics—the first time this had ever happened.

Tennis

✎ Czech-born American Martina Navratilova won the Women's singles at Wimbledon a record nine times.

✎ Navratilova won 167 singles titles in her career—more than any other player ever.

✎ American Pete Sampras won a record 14 Grand Slam titles.

✎ Australian Rod Laver twice won four Grand Slam titles in a single year.

Soccer

➤➤ Brazilian striker Pele scored an average of one goal every time he played for Brazil. He hit 90 hattricks in his career.

➤➤ While Pele played for them, Brazil won the World Cup three times.

➤➤ Ukrainian Nikolai Kutsenko once played "keepy-uppy" for $24\frac{1}{2}$ hours.

➤➤ U.S. women's soccer player Mia Hamm has scored more goals in international football than any other player, man or woman: 149.

Tiger the Driver

☞ Tiger Woods first played golf on TV at the age of just two.

☞ Woods became the youngest world number one golfer ever at the age of 21 in 1996.

☞ By 2006, Tiger Woods had won 12 major golf championships.

☞ By 2006, Woods had notched up 52 wins on the PGA tour, more than any other golfer.

☞ In 2005, Woods was the highest paid professional athlete in the world.

THE HUMAN WORLD

In our relatively brief time on the planet, humans have done some amazing things. We've built many machines, including computers that could soon be as powerful as our brains. We've also done some silly and some terrible things. Here's a selection of the best and the worst of the human world.

Brilliant Buildings!

Data Bites!

Astounding Autos!

Simmering Cities

Over 3 billion of us now live in them, and thousands more flock to the world's cities every day.

Purpose-built Cities

➤➤ Russian Czar Peter the Great chose a swamp for his new capital city, St. Petersburg. Over 30,000 builders died in the perilous conditions.

➤➤ Brazil's new capital city, Brasilia, was the brainchild of President Juscelino Kubitschek, who wanted a capital in the middle of the country. It took 41 months to build and was finished in April 1960.

Ancient Cities

☛ The world's first city may have been Hamoukar in Syria, dating back over 6,000 years. However, no one lives there anymore—unlike nearby Damascus, where people have lived continuously for over 10,000 years.

☛ Babylon, near modern-day Baghdad, was the first city with more than 200,000 people, back in 600 B.C.

☛ Rome was the first city to reach 1 million, in A.D. 27.

☛ The first big city in America was Teotihuacán in modern Mexico, where 250,000 people lived in A.D. 300.

➤➤ Rome's Vatican City, left, is called a city, but it is actually a country. It's also only the size of 50 soccer pitches!

➤➤ With 329 citizens, Fürstenau in Switzerland is the world's smallest city by population.

HONOREM PRINCIPIS APOST PAVLVS V BVRGHIVS ROMANVS PONT MAX AN MDCXII PONT VII

TRUE MEGA CITY FACTS

☛ The biggest cities in the world are called megacities. The biggest in area is New York. By population, the biggest is Tokyo.

☛ In 2006, five cities in the world were home to over 18 million people – Tokyo (35.5), Mexico City (19.24), Mumbai (18.84), New York (18.65), and São Paulo (18.61).

☛ Five of the world's biggest 13 cities are on the Indian subcontinent: Mumbai, Delhi, Calcutta, Dhaka, and Karachi.

☛ The most crowded megacity is Manila in the Philippines, – one and a half times as many people pack into every square mile as they do in Los Angeles!

➤➤ For many years, Tokyo (pictured) was the most expensive city in the world to live in. Now Moscow and London cost even more.

➤➤ Three times as many people live in every square mile in Tokyo as in the same area in London.

Modern cities have grown so much that they have swallowed up small towns around them. If you only count cities' original areas, Mumbai in India is the biggest of all.

Cities of the Future

☛ By 2020, Lagos, Dhaka, Delhi, Mumbai, Calcutta, Jakarta, São Paulo, and Mexico City will each be home to over 20 million people.

☛ The biggest megacity may one day be China's Pearl River (PRD), centered on Guangzhou.

☛ Between now and 2020, we will see the biggest migration in human history, as half a billion people move to China's rapidly growing cities.

Fantastic Buildings

We put up all kinds of buildings to live and work in, and build them using bricks, stone, even dung.

Ancient Edifices

➤➤ In 2000, the remains of the world's oldest known building, half a million years old, were found in Japan.

➤➤ Built 4,500 years ago, Egypt's Great Pyramid is still the world's biggest building, with 116.5 million cubic feet of stone—enough to build a 3-foot-high wall around France!

➤➤ The world's first large concrete building was not a modern skyscraper but the Pantheon in Rome, built in A.D. 126.

In places like this in Turkey, people called troglodytes live in caves dug out of the rock. Caves may look primitive, but they're nice and cool in summer and warm in winter.

Dung Houses

☛ The stones in this Rwandan rondavel are stuck together using cow dung. Dung mixed with straw also makes great bricks.

☛ Cow dung may be the building material of the future, used as both bricks and mortar.

☞ A vital development for skyscrapers was the invention of the elevator by Elisha Otis in 1857—think of all those stairs without it!

☞ The world's first skyscraper may have been the Home Insurance Building in Chicago, built in 1885. But New York was soon out-scraping Chicago, with buildings like the Flat Iron (1902).

☞ For decades, the Empire State Building, built in New York in 1931, was the world's tallest building, standing at 1,250 feet.

☞ The tallest building in the world is now Taiwan's Taipei 101, pictured right. It was completed in 2004 and towers 1,700 feet high.

☞ Not far behind are the twin towers of the Petronas Building in Kuala Lumpur in Malaysia, built in 1998, which soar 1,480 feet into the air.

The Future

✎ More houses will be designed to minimize energy use, with windmills and solar panels, and all kinds of devices for recycling waste, even ways to reuse feces!

✎ In "intelligent" houses, things such as the lights and heating will be electronically controlled by your computer.

✎ Your refrigerator will read bar codes on food and tell you when the food's out of date—and maybe even order more from the supermarket.

✎ Cleaning jobs will be done with sound waves, not soap and water. You may even shower using sound.

Terrific Tools

We've come a long way from the first people who thought to pick up a stone and sharpen it.

Stone Tools

➻ Human ancestors learned to use stones as tools about 2.6 million years ago. The age of stone tools—the Stone Age—lasted over 2.59 million years.

➻ Shaping a stone tool to give it a sharp cutting edge is called "knapping," and in the Stone Age there were factories where a load of knappers would often be caught knapping.

The First Machines

☞ Stone Age hunters invented ways to increase their muscle power when hunting, such as sling arms to throw spears and bows to fire arrows. These were the first machines.

☞ 12,000 years ago, people began to farm, and invented machines, such as plows, to help them in their work.

☞ 7,000 years ago, someone in Sumer, in the Middle East, turned a few potters' wheels on their side and added them to a cart to invent the wheel. That wheel was fundamentally the same as those we use today.

People in Mesopotamia realized they could take the sweat out of plowing 8,000 years ago by getting an ox to do the hard work. The ox was the first engine.

Basic Machines

✏️ Scientists say there are just six "simple" machines on which all others are based: lever, ramp, wedge, screw, wheel or roller, and pulley. We all still use these basic machines in our homes.

✏️ Machines often give you greater power by making you move further. For instance, using pulleys you can lift an object by one yard while pulling a rope ten yards. This means you can lift an object ten times heavier than you could without the machine.

Animal Tools

☛ Humans used to think the Stone Age was what made us special. But chimps have had a Stone Age, too. They've been using stone tools to crack nuts for at least 4,000 years.

☛ Egyptian vultures use stones to crack tough eggs.

☛ Crows on the Pacific Island of New Caledonia shape special sticks to get insects out of dead wood.

☛ In 1980, when American scientist Ben Beck forgot to add water to the dried mash he fed his laboratory crow, the crow fetched his own water in a cup.

TRUE NANO FACTS

☛ Scientists are working to develop nano-tools—tools a million times smaller than a pinhead. (A "Nano" means one-billionth of a yard.)

☛ Molecular gears such as these, right, could be used to make machines many times thinner than one strand of human hair.

☛ Nano-vehicles could travel inside your body to deliver drugs directly to where they are needed.

I've Got an Idea!

There have been some brilliant inventions, and some very silly ones.

Oh Dear...

✎ Fed up with waiting for his luggage at airports, in 1945 Robert Fulton invented an airplane that converted into a car. Unfortunately, it was so heavy it could never take off.

✎ One inventor has a patent for coffins built with alarms, in case someone is accidentally buried alive.

✎ Another inventor has a patent for two-person gloves, special gloves for couples who like to hold hands skin-to-skin in icy weather.

✎ Thomas Alva Edison, who helped invent the light bulb and movies, also invented concrete furniture. It went down like a concrete balloon.

Life-changing Drugs

☛ Alexander Fleming, left, got the idea for antibiotic drugs when he noticed how the fungus penicillin killed bacteria he was studying in his laboratory.

☛ In 1952, Jonas Salk developed a vaccine against polio. He tested it on himself to be sure it was safe.

The whoopee cushion was invented when workers at the Jem Rubber Company in Toronto, Canada, "experimented" with rubber sheets.

The fart alarm is a joke invention that is said to beep when it detects a fart. In fact, it is set off by vibration.

Roller Skates

Roller skates were invented in 1760 by London inventor Joseph Merlin. One day, Merlin made a spectacular entrance to a fancy dress ball wearing metal-wheeled boots—and crashed headlong into a giant mirror, shattering it.

In the 1970s, Art Fry realized he could use a glue that didn't stick to make a bookmark in his hymnbook—and invented the Post-it note.

TRUE INSPIRED FACTS

In 1956, Noah and Joe McVicker invented a wallpaper cleaner. They then realized it was like modeling clay only better, and so invented Play-doh.

Roy Plunkett was working with gases in 1938 when one turned into a solid with a surface so slippy nothing stuck to it. He had discovered Teflon, which is used to make nonstick pans.

In 1905, 11-year-old Frank Epperson accidentally left out a cup of soda water with a stick in it on a cold night—the first popsicle.

Inspired by the reflection of lights on road signs, Percy Shaw invented the studs now set in unlit British roads, called catseyes, in 1934.

Made It!

The first modern factories were started to make cloth. Now we make all kinds of things in them.

Industrial Revolution

☞ In the 1700s, the Industrial Revolution began in Britain when cotton was turned into cloth on machines in factories.

☞ By the late 1800s, many companies, such as Singer with its sewing machine, right, were making machines that you could use at home to make clothes.

With millions of men away fighting, women worked in the arms factories during World War I. In many cases, they were working outside the home for the first time.

TRUE FACTORY FACTS

☞ Henry Ford got the idea for the assembly line to make the Model T Ford, the world's first cheap, mass-produced car, from a visit to a slaughterhouse, where he saw the carcasses of cows swinging along in a line.

☞ Making cars is now the world's largest manufacturing industry, employing tens of millions of people to make well over 60 million new cars and trucks each year.

☞ The world's biggest factory is Boeing's Everett Plant, built to make the 747 Jumbo Jet. The factory covers an area the size of 40 football pitches!

Filthy Factories

➤➤ The coming of factories upset many people, because they put many craftsmen out of work. People called Luddites smashed the machines until the army moved in to stop them.

➤➤ The world's dirtiest factory is probably in Rannipet in India, where a chemical factory has built up a pile of 1.4 million tons of toxic waste.

➤➤ In the city of Dzerzinsk, Russia, the local weapons factory has polluted the water so much that men on average die at the age of 42.

➤➤ The world's biggest chocolate factory is the Hershey factory in Pennsylvania.

➤➤ The Hershey factory gets through over half a million pounds of cocoa beans in a year.

➤➤ A robot the size of a person can move a load of 110 pounds to an accuracy of 0.004 inches.

➤➤ Modern factory robots can perform even skilled tasks like welding much faster and more accurately than human welders.

Fabulous Food

We can grow all kinds of food on farms, but if we're not careful, we can end up eating too much of it.

The First Farmers

☞ People started to farm 12,000 years ago. They selected seeds from wild grasses and planted them to grow cereals such as wheat.

☞ The new farmers were shorter than those who hunted and gathered fruit and nuts, because their diet was less varied.

➤➤ A century ago, one farmer in the United States could grow enough grain to feed 25 people. Using machines, one farmer can now grow enough to feed 1,000.

➤➤ In the future, harvesting machines such as this will be controlled by computers.

Farm Animals

☞ Today we share the world with over 1.8 billion sheep, 1.3 billion cows, 1 billion pigs, and over 13 billion chickens.

☞ In Georgia, Arkansas, and Alabama combined, 4 billion chickens are killed every year.

☞ In the industrialized world, few farm animals now eat only grass. They are also fed grain, soy, and manure.

☞ Over a billion people in the world are overweight from eating too much food. About the same number are ill or dying from too little.

☞ One in three children born in the United States in 2000 is predicted to become diabetic—probably because of excess sugar in their diets.

☞ Children in the United States and Europe eat more than twice the recommended amount of salt in their diet every day.

☞ Three-fourths of the salt in our diets comes from processed food. Extra salt is added to it.

None of these fruit candies contains any fruit. Instead, like most processed foods, their flavors are achieved using cocktails of flavor chemicals created in laboratories.

➤➤ Greenhouses, freezing techniques, and transport used today ensures we get almost any food all year.

➤➤ The contents of the average shopping cart travel 93,000 miles to get there.

Food for Thought

☞ One-third of all food produced in Great Britain is simply thrown away.

☞ One-third of all the fruit and vegetables we eat contains traces of the chemical pesticides sprayed on the fields and plants they came from.

☞ It takes 1,000 gallons of water to make 2 pounds of cheese, and over 22,000 gallons to produce 2 pounds of hamburger beef.

Toilet Humor

Going to the bathroom can be a serious business. What exactly do you do with all the poop?

Medieval Toilets

☛ In medieval castles, special rooms poked out from the walls. Inside was a wooden board with a hole in it to plant yourself on. The poop fell through the hole into a pit.

☛ The castle's toilet was called a garderobe because the lord of the castle kept his clothes there (it's French for "guard robes"). The horrible smell kept moths away.

☛ The pit where the waste from a garderobe collected was cleared up by a lucky man called a gongfarmer.

Ancient Toilets

✎ The world's oldest known toilets are in the 4,000-year-old Palace of Knossos on Crete. Pipes carried waste away and an overhead tank provided water for flushing. Very sophisticated!

✎ Romans shared a stick with a sponge on it to wipe themselves. You didn't want the wrong end of the stick!

In Georgian times, 200 years ago, men would stop mid-conversation to grab a potty like this one to pee. Ladies would discreetly place it under their skirts.

☛ In Victorian London, human sewage drained into the Thames River, turning drinking water brown.

☛ The water people drank was so unhealthy that in the 1830s and 1850s thousands of people caught the disease cholera from it and died.

☛ In 1848, the government decreed that every new house should have an ash-pit—where poop and pee fell onto a pile of ash ready to be carted away by the "night soil man."

☛ In 1858 a heat wave caused the "Big Stink." London smelled like one, big, recently used toilet!

☛ Finally, an underground sewer system was built. It was finished in 1865 and is still in use today.

Toilet Paper

➤➤ People used sponges, rags, leaves, and even their hands to wipe their buttocks until 1857, when toilet paper was introduced. It was sold under the counter in chemists and was called "curl paper," because girls had used twists of paper like it to curl their hair.

➤➤ Toilet rolls were first introduced in 1928.

Toilets of the Future
Some Japanese toilets have lids that rise when you approach, seats that warm as you sit down, and give a squirt of water to wash you when you're done.

Data Bites

We all like machines to do our adding for us. Modern computers can do a whole lot more.

TRUE COMPUTER FACTS

☛ Possibly the first ever computer was made over 2,000 years ago in Crete. It used a complex mechanism of wheels and gears to compute the movements of the stars.

☛ In 1832, long before the days of electronics, Charles Babbage built a calculator that used brass wheels and dials, called the Difference Engine (right).

☛ Modern computing began in 1936 when Alan Turing developed the idea of sets of instructions for the computer called algorithms.

Computers contain chips made of silicon, such as this one. Each chip is made of thousands of switches, which can either be 0 or 1. Everything a computer does is done using just 0s and 1s.

Computer Power

☛ A teraflop sounds like a useless horror film. It's actually a term that means a trillion calculations a second, used to measure computer power.

☛ The world's most powerful computer is the Blue Gene/L, which may be able to reach 360 teraflops. That's still less than the human brain, which works at 10,000 teraflops.

☛ They call Blue Gene compact, because it's only half as big as a tennis court!

☛ IBM is building Blue Gene/P, which will be able to do one petaflop— one quadrillion calculations per second.

In a famous challenge in 1997, the chess-playing computer system known as Deep Blue beat world chess champion Garry Kasparov in a chess match.

Internet

☛ It took 13 years for television to reach 50 million users. It took the Internet less than four years. As of now, there are over 260 million people with Internet access worldwide.

☛ An e-mail address receives an average of 26 e-mails a day.

POWER
DS
US
ONLINE
LINK
TEL 1
TEL 2
STANDBY

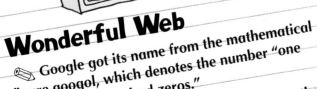

Wonderful Web

✎ Google got its name from the mathematical figure googol, which denotes the number "one followed by a hundred zeros."

✎ Yahoo! is named after characters in Jonathan Swift's book *Gulliver's Travels*. A Yahoo is a horrible person who is barely human!

Ships Ahoy!

Vikings first crossed an ocean 1,000 years ago. Now huge ships carry goods right around the world.

Walking the Plank

A popular image of pirates is that they made people walk along a plank and jump into the sea, to be eaten by sharks. It isn't true. Author J.M. Barrie made it up for his book *Peter Pan*.

TRUE SOS FACTS

☞ In the 1890s, ships started to carry radios. They could talk to each other with machines such as this one using a system of clicks called Morse Code.

☞ A ship's distress call is Morse Code for SOS. This was chosen in 1908 because the letters were easy to remember and key in to radios. It's a myth that SOS stands for "Save Our Souls."

☞ Speaking on the radio, the term "May Day" is used to signal for help. It comes from the French term "m'aidez," which means "help me."

Cruise News

➽ **The *Queen Mary II*** is the largest ocean liner ever. It weighs 136,000 tons and features no fewer than 15 restaurants and 5 swimming pools.

➽ ***Freedom of the Seas*** can carry more people than any other ship. It can carry 5,600 people—4,300 passengers and a crew of 1,300.

➽ **The *Queen Elizabeth II*** moves only 6 inches for each gallon of diesel fuel that it burns.

➽ When launched, the ***Titanic*** was the largest ship ever, weighing 42,000 tons, but it sank on its first voyage when it hit an iceberg on April 14, 1912.

Speedy Boats

☛ Donald Campbell set seven world water speed records. he was killed in 1967 while going for his eighth.

☛ The world water speed record of 317 mph was set in 1978 by Ken Warby on Blowering Dam in Australia.

➽ **Many supertankers** are so long you could fit the Eiffel Tower laid on its side on their decks.

➽ **The biggest** is the *Knock Nevis*. It weighs 494,100 tons empty and 748,832 tons when full of oil. It's 1,500 feet long.

Flying High

We spent a long time trying, and failing, to imitate the birds. Now we're finally getting it right.

Airships

In the 1930s luxury airships filled with hydrogen cruised across the Atlantic. They were stopped after the airship *Hindenburg* caught fire in 1937. Recently airships have been built filled with safe helium.

Early Flight

➤➤ Early attempts at flight involved men trying to beat gravity by strapping wings to their arms. Gravity always won.

➤➤ George Cayley built a glider in which his poor butler became the first aviator, flying across the Yorkshire Dales in 1853.

➤➤ German Otto Lilienthal, with a hang-glider-like plane, was the first man to make repeated, controlled flights in the 1890s. He died in 1896 when a gust of wind broke his wing.

➤➤ Americans Orville and Wilbur Wright made the first successful powered flight in 1903.

TRUE DOG FIGHT FACTS

☛ The first aerial battles in World War I came to be called "dogfights" because each plane tried to get behind the other, like two dogs chasing each other.

☛ The greatest fighter ace was German Manfred von Richtofen, known as the "Red Baron," who won 80 air battles before being shot down in 1918.

☛ According to the pop song "Snoopy vs the Red Baron," he was shot down by cartoon dog Snoopy!

➤➤ First flown in 1969, Concorde was a supersonic plane—that is, able to fly faster than the speed of sound.

➤➤ Concorde once flew from New York to London in 2 hours 52 minutes and 59 seconds.

➤➤ The Harrier Jump Jet can take off vertically (i.e. without needing a runway).

➤➤ The fuel in a Jumbo Jet would take a car around the world four times.

Stealth Planes

☞ Stealth aircraft use special design features to avoid being detected by radar, such as a shape that deflects the radar waves or matt paint, which absorbs them.

☞ The Lockheed Skunk Work's *Cormorant* stealth aircraft starts and ends its flight hidden 165 feet underwater!

Future Flights

✎ In March 2004, NASA's experimental plane the *Hyper X*, powered by a "ramjet," flew at 5,000 mph—over seven times the speed of sound.

✎ Modified ramjets, called scramjets, might one day power passenger planes that can fly from London to Sydney in just 90 minutes.

Railroad Roundup

When coaches were pulled by horses, journeys were slow and bumpy. Trains brought new levels of speed and comfort.

Early Trains

➤➤ The world's first railroad, called the Diolkos, was built in 550 B.C. in Greece. Coaches were pulled by slaves!

➤➤ The first successful railroad engine was the steam engine built by English engineer Richard Trevithick in 1804.

➤➤ The first railroad death was that of William Huskisson, on the day the first passenger railway opened in 1830.

TRUE TRAIN FACTS

☞ George Stephenson's *Rocket*, below, won the first locomotive trials in England in 1829, averaging 30 mph over a 50-mile track.

☞ By the 1840s, trains could complete the 185-mile journey from London to Exeter, south of England, in under four hours.

☞ On May 10, 1869, railroads from either side of the United States met at Promontory, Utah. This completed the first railroad ever to cross a continent.

Railroad Records

☞ The world speed record for a steam train was set by the *Mallard*, which pulled seven coaches at 125 mph on July 3, 1938.

☞ The longest train ever was 4½ miles long and contained 660 coaches. It traveled from Saldanha to Sishen in South Africa on August 26, 1989.

☞ The fastest train ever was the French TGV, which whizzed between Courtalain and Tours at 320 mph on May 18, 1990.

☞ The Trans-Siberian Express is one of the world's longest train services, taking seven days to cover 5,771 miles from Moscow to Vladivostok in eastern Siberia.

☞ The world's fastest regular train service is Japan's Shinkansen bullet train which covers the 119 miles from Hiroshima to Kokura in just 44 minutes, at an average speed of 163 mph.

London's underground railroad system is the oldest and largest in the world. It began with the Metropolitan Railway (pictured here) on January 10, 1863.

➡ Japan's Shinkansen bullet trains, left, are shaped like airplanes to reduce air resistance.

➡ New Maglev trains glide not on wheels but on a magnetic cushion.

Astounding Autos

The car brought us the freedom to travel whenever we wanted—when we're not stuck in a traffic jam, that is.

Mrs. Benz's Hairpin

When German Karl Benz built the world's first gas-engined car in 1885, his wife Berta "borrowed" it to take their sons on a trip. She used a hairpin to keep it going after it broke down.

The long roads across Australia allow a single truck to pull several trailers in a "road-train." The longest road-train ever had **79** trailers.

Best Sellers

☞ The first mass-produced car, in 1905, was Ford's Model T. Over 15 million were sold.

☞ The best-selling car is the VW Beetle. Over 22 million have been sold.

Fastest Cars

➤➤ **Thrust SSC** is the world's fastest car. It broke the land speed record at 763 mph on October 15, 1997, driven by Andy Green at Black Rock Desert, Nevada.

➤➤ **Thrust's** record-breaking run was the first time a car had traveled faster than the speed of sound.

Weird and Wonderful

✎ The lightest car ever was the Suminoe Flying Feather of 1954. It weighed just 997 pounds.

✎ The DUKW of 1942, known as the Duck, was the most popular amphibious vehicle ever. Amphibious vehicles are able to travel on the road and in water.

Dynasphere

One of the oddest cars ever was the Purves Dynasphere, which looked just like a single giant tire, in which the driver sat in the middle.

Fastest Road Cars

☞ The Bugatti Veyron, left, is the world's fastest production car, reaching a top speed of 253 mph.

☞ The Barabus TKR is the fastest road car. It can reach 267 mph.

☞ Another speedster, the Saleen S7 Twin Turbo, can reach 258 mph.

☞ None of these fast cars costs less than $375,000. And they cannot be driven at even half their top speed on most roads.

EARTH
AND SPACE

Our planet can feel like a pretty extreme place at times. In Arica in Chile, it barely rains enough in a century to fill a teacup, and on Mount Everest, 24,000 feet up, the air is too thin to breathe. But that's nothing compared to the 200-year storm that's been raging on the planet Jupiter!

Race to Space!

Violent Earth!

Desolate Deserts!

Awesome Oceans

The Earth is known as the Blue Planet because nearly three-quarters of its surface is covered in water.

Wealthy Water

Dissolved in every cubic ¼ mile of seawater is about 13 pounds of gold. The oceans contain more than 240 million cubic miles of water, which means there are 5 ½ million tons of gold in the Earth's oceans!

TRUE OCEAN FACTS

☞ The longest mountain range on the Earth is the Mid-Ocean Ridge, which stretches 46,000 miles, more than four times the lengths of the Andes, Rockies and Himalayas combined.

☞ If the oceans' total salt content were dried, it would cover all land on the Earth to a depth of 6 ½ feet.

☞ The deepest point in the ocean is Challenger Deep, in the Mariana Trench, 11,947 yards down in the Pacific Ocean.

☞ The temperature of most ocean water is about 39 °F, just above freezing.

☞ Rip currents are the oceans' biggest killer of humans. They happen when the sea is twisted as waves break differently on different parts of the shore.

Steamy Seas

Recent studies suggest the Atlantic was like a steaming hot tub between 84 and 100 million years ago, with temperatures that ranged between 91 °F and 108 °F.

The oceans are a hotbed of volcanic activity. In fact, 90 percent of all volcanoes are under the sea. The world's biggest concentration of active volcanoes is in the South Pacific, where there are over 1,100 active volcanoes.

The world's tallest mountain is actually in the sea. It's the volcanic island Mauna Kea in Hawaii, and is 33,465 feet tall from the ocean floor to the summit.

The Oceans Deep

☞ Under the Denmark Strait, a waterfall slowly cascades downward for over $1^3/_4$ miles, over three times as far as Angel Falls, in Venezuela, which is the tallest land waterfall.

☞ For years, scientists wondered what creatures on the seabed ate. The puzzle was recently solved by the discovery of snotlike "sinkers" made by tiny tadpole-like creatures, which double the available food.

☞ Some bugs can feed entirely on chemicals spewed up by seabed volcanoes. The bugs provide food for other deep-sea creatures, such as giant clams.

The Ocean Glows

Everyone used to think sailors had drunk too much rum when they talked about the ocean glowing at night. Now satellite pictures have shown that large patches of the Indian Ocean may glow two or three nights in a row. Scientists suspect the cause is giant colonies of luminous bacteria, which sometimes light up waves such as this one.

Desolate Deserts

There are places on the Earth where it hardly ever rains, and they can be searingly hot or unbearably cold.

TRUE SAHARA FACTS

☞ The world's biggest desert by far is the Sahara in Africa. The Sahara is bigger than the four next largest deserts put together. It's pretty much as big as the whole of Europe.

☞ Barely 10,000 years ago, parts of the Sahara were green, and it even used to snow in some places.

☞ However, there are places in western Sahara where there have been only very small amounts of rainfall for well over 2 million years.

☞ All those dry years have created an area of sand dunes over 980 feet high called the erg. The world's biggest sand dune, which is over 1,640 feet high, is found in the erg.

☞ The dunes move across the desert like very slow waves, about 3 feet every year.

What's a Desert?

➤➤ There are different measures of what makes a desert, but it is usually agreed that a true desert is an area that gets less than 10 inches of rain a year.

➤➤ Deserts cover about one fourth of the Earth's land surface.

Dew Slurper

Australia's deserts are home to the thorny devil lizard. It can survive without rain by drinking the dew it collects in its spikes.

All Dried Out

➤➤ The world's driest place is Arica, in Chile, where there is barely enough rain in a century to fill a teacup. Arica's annual rainfall is just $^3/_{100}$ inch.

➤➤ Although there's plenty of ice, Antarctica is actually the driest continent on the Earth because all the water is frozen. It's also the windiest continent—and of course the coldest!

The flowering shrubs called creosote bushes (above), in the Mojave Desert, are among the oldest living things on the Earth at 12,000 years old.

Lost in the Fog

In the Namib Desert in Namibia, southwest Africa, it rarely rains, but it is often very foggy. The fog comes from moisture drifting in off the Atlantic Ocean.

Searing heat and lack of rain make California's Death Valley an appropriate name. On July 10, 1913, temperatures there reached 135 °F—among the highest ever recorded.

Mountains

As the earth's surface buckles and crumples, great mountains are pushed into the skies.

TRUE EVEREST FACTS

☞ The world's highest mountain is Mount Everest in Nepal.

☞ The height of mountains used to be measured from the ground using spirit levels and sighting devices to measure angles. Now mountains are measured more accurately using satellite techniques. Satellite measurements in 1999 raised the height of Everest from 29,029 feet to 29,035 feet.

☞ By the end of 2004, 2,250 climbers had succeeded in reaching the top of Everest—but 186 had died in the attempt on the mountain.

☞ On May 5, 1973, a 16-year-old Nepalese boy Shambu Tamang became the youngest person ever to climb Everest.

☞ The first married couple ever to climb Everest was Andrej and Mariga Stremfelj from Slovenia. They reached the summit on October 7, 1990.

More than 100 people have climbed the highest peaks on each of the seven continents, including the highest and hardest of all, Everest (above).

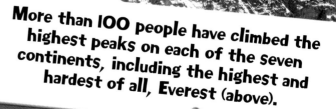

Temperatures drop 1 °F every 328 feet you climb, so mountain peaks are very cold. The air is thinner on mountains, so climbers may also need oxygen masks to breathe.

A Room With a View

☛ The Japanese-run Everest View Hotel is, at 12,780 feet, the highest hotel in the world. Guests fly to an airstrip in the Khumbu region of the Himalayas and are then transported by yak to the luxury hotel.

☛ Because the air is so thin this high up, the hotel pumps extra oxygen into each room to stop guests from getting altitude sickness.

Rain and snow falling on mountains is the source of over half the world's fresh water, and all the world's major rivers are fed mainly by mountains.

Mount Up

➤➤ The rock in mountains is not completely rigid. It flows like thick mud, only much more slowly, over millions of years. Ranges such as the Himalayas are flowing out and getting flatter at the edges.

➤➤ All mountain ranges tend to start flat. Mountains can build up where the movement of the Earth's crust tilts, crumples, squeezes, and lifts these flat layers.

➤➤ High ranges are geologically young because they are soon worn down. The Himalayas are 25 million years old—relatively young.

➤➤ The central peaks of the Andes and Himalayas are rising by about ½ inch per year.

The World's Highest

✎ **Highest Mountain in Africa**
Kilimanjaro, Tanzania: 19,340 feet

✎ **Highest Mountain in Antarctica**
Vinson Massif: 16,066 feet

✎ **Highest Mountain in Asia**
Everest, Nepal: 29,035 feet

✎ **Highest Mountain in Australia**
Kosciusko: 7,310 feet

✎ **Highest Mountain in Europe**
Elbrus, Russia (Caucasus): 18,510 feet

✎ **Highest Mountain in Western Europe**
Mont Blanc, France-Italy: 15,771 feet

✎ **Highest Mountain in North America**
McKinley (Denali), Alaska: 20,321 feet

✎ **Highest Mountain in South America**
Aconcagua, Argentina: 22,841 feet

Raging Rivers

Water that falls as rain on the ground drains into rivers, which carry it down to the oceans.

TRUE AMAZON FACTS

☞ The Amazon River is the source of about one-fifth of all the water that floods into the oceans from rivers.

☞ The Amazon in flood could fill the world's biggest sport stadium with water in 13 seconds.

☞ Many of the Amazon's tributaries are themselves major rivers. Seventeen of them are over 1,000 miles long!

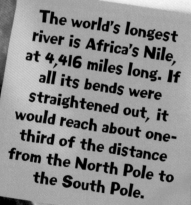

The world's longest river is Africa's Nile, at 4,416 miles long. If all its bends were straightened out, it would reach about one-third of the distance from the North Pole to the South Pole.

River Run-down

✐ By most measures, the Amazon is the world's second longest river at 3,963 miles long. The third longest river is China's Yangtse—at 3,915 miles.

✐ The muddiest river is China's Yellow River. Two billion tons of yellow mud wash down it each year.

✐ The world's longest tributary is the Madeira (2,100 miles), which flows into the Amazon.

✐ The area where a river's freshwater mixes with the sea's saltwater is called an estuary. The world's longest estuary is the Ob in Russia—up to 50 miles wide and 550 miles long.

Waterfalls

➤ At Victoria in Zimbabwe, the Zambesi River pours over a cliff ²/₃ mile wide and 350 feet high, making a roar that can be heard over 25 miles away.

➤ The world's highest falls, Angel Falls in Venezuela, plunge 3,212 feet.

Iguazu Falls

At times, over 1.4 million gallons of water pour every second over the 275 branches of the Iguazu Falls in Brazil, a peak flow second only to Zimbabwe's Victoria.

The World's Water

☞ There are 434 million cubic miles of water in the world.

☞ 315 million cubic miles (93 percent) of this is seawater.

☞ 9 million cubic miles (2.5 percent) is in aquifers deep below the Earth's surface.

☞ 7 million cubic miles (2 percent) is frozen in polar ice caps.

☞ 55,000 cubic miles of water passes through lakes and rivers.

☞ 4,000 cubic miles of water is atmospheric moisture.

☞ 3,360 cubic miles of water is inside the bodies of living things.

Of the world's 500 largest rivers, over half are seriously polluted and many are disrupted by dams such as the Hoover Dam in the United States (above), which threaten fish.

Violent Earth

The ground may feel solid, but it's just a thin crust floating on top of the Earth's liquid core.

Major earthquakes that cause damage, such as this, are set off by the movement of tectonic plates—the 20 or so giant slabs of rock that make up the Earth's surface.

Quakey-quakey

✎ Most earthquakes last less than a minute. The longest recorded, in Alaska on March 21, 1964, lasted just four minutes.

✎ There are half a million detectable earthquakes each year.

✎ China has come up with an earthquake prediction system that relies on the behavior of snakes. They leave their nests just before a quake, even in winter.

Tectonic plates normally slide by $1\frac{1}{5} - 1\frac{3}{5}$ inches each year. In a major quake, they can jump 3 feet or more. Sometimes they pull away from each other, causing gaps like this.

After a large earthquake, the Earth rings like a bell, but at a very, very low pitch.

The world's deadliest recorded earthquake occurred in 1557 in China, killing over 800,000 people.

An earthquake in 1811 sent the Mississippi River flowing backwards.

Lava Lowdown

☞ The temperature of lavas varies depending on its chemical composition. Hawaiian lava (basalt) is usually around 2,000 °F.

☞ Lava fountains can propel lava up to 2,000 feet above a volcano. They are driven by bubbles that form from gas dissolved in the lava.

TRUE VOLCANO FACTS

☞ Most volcanoes occur near cracks between the tectonic plates that make up the Earth's surface.

☞ The eruption of Mt. Tambora in Java in 1815 sent up so much ash that the sun was blocked out around the world, giving two years of cool summers.

☞ The eruption of the volcanic island of Krakatoa near Java in 1883 could be heard one-fourth of the way around the world!

☞ One of the most gigantic eruptions ever occurred in Yellowstone 2.2 million years ago. It poured out enough magma to build the biggest current volcano in the world six times over.

☞ There are around 1,510 active volcanoes in the world.

☞ Cuexcomate in Mexico (43 feet) is the world's smallest volcano.

Rock Balls

Mercury, Venus, and Mars have a rocky crust around them like the Earth.

Beautiful Venus

☛ Wrapped in yellow clouds, Venus looks attractive, which is why it was named after the Roman goddess of love. But those gorgeous clouds are made of sulfuric acid!

☛ Venus is the hottest planet in the Solar System, with a surface temperature of 880 °F.

More on Venus

✎ Venus's atmosphere is so thick that the pressure on the surface is 90 times that of Earth—enough to crush a car completely flat.

✎ Venus turns very slowly. It takes about 225 days to go around the Sun, yet it takes 243 days to turn on its axis. So a Venusian day is longer than its year.

✎ Unlike most of the planets, Venus turns in the opposite direction to its orbit. Scientists still don't understand why this happens.

✎ The clouds on Venus sometimes glow with lightning. The astronomer Franz von Gruithuisen imagined the glow to be fires lit by Venusians for the coronation of a new emperor.

Venus's clouds are many miles thick. The image on the left shows the clouds at the surface, on the right those several miles underneath.

Speedy Mercury

☛ You could go skiing on Mercury because it may well have polar ice caps, but you'd need a special ski outfit because the ice is made from pure acid.

☛ Mercury is so close to the Sun it gets right around in just 88 days (compared to 365 days for the Earth). Yet it spins very slowly, taking over 58 Earth days. So there are fewer than two days in Mercury's year. That means you'd get a birthday every other day!

☛ The ancient Romans called the planet Mercury after the fleet-footed messenger to the gods in their myths.

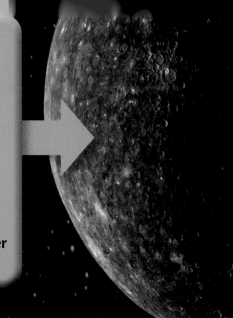

Mars, the Red Planet

☛ Mars (left) is named after the Roman god of war because it sometimes appears a blood-red color.

☛ Because there are no oceans on Mars, the area of land on the planet is much the same as all the continents on the Earth combined. So studying all the mountains, valleys, and plains on Mars is as big a task as studying all those on the Earth.

☛ The longest canyon in the Solar System is Mars' *Valles Marineris*. It is more than 3,100 miles long, so it would stretch from California to New York.

➤➤ Mars has the largest dust storms in the Solar System. Sometimes they can cover the entire planet (far right).

➤➤ Mars has ice caps at each pole, which are made of both ice and solid carbon dioxide (dry ice).

Gas Giants

Four planets far from the Sun are great big balls of gas.

☛ Jupiter is the biggest planet in the Solar System by far.

☛ Jupiter may be big but it spins at incredible speed. It turns right round in less than 10 hours—compared with 24 for the Earth. Since the planet is 280,000 miles around the middle, this means the surface is whizzing round at nearly 30,000 miles an hour! Hold on tight!

☛ Jupiter is so massive that it is squeezed by its own gravity enough to make it glow under the pressure.

☛ Jupiter's Great Red Spot is a giant storm that has been raging for over 200 years.

Huge Planets

✎ Beneath Jupiter's thin atmosphere may be an incredibly deep ocean—15,500 miles deep! That's more than enough to drown Earth in!

✎ Because Jupiter is mostly liquid, its fast spin makes it bulge in the middle.

✎ All four gas giants, Jupiter, Saturn, Neptune, and Uranus are thought to have a small solid core made of rock or metal.

✎ Uranus and Neptune are also called the ice giants. They are so far from the Sun that they are mostly frozen inside.

Icy Uranus and Neptune

☞ Uranus (left) and Neptune (below) are so far from the Sun that surface temperatures are -346 °F!

☞ Uranus and Neptune are both blue because of the blue gas methane in their atmospheres.

☞ In summer on Uranus, the Sun does not go down for 20 years! It just keeps on going around and around in the sky.

☞ In winter, it's dark on Uranus for 20 years!

☞ Neptune has the strongest winds of any planet in the Solar System. Gusts can reach over 1,250 miles per hour.

Sleek Saturn

☞ Saturn is so light that if you could find a bath big enough to put it in, it would actually float.

☞ Saturn is surrounded by a halo of rings that stretch over 45,500 miles out from the planet but are less than $2/3$ mile thick. They are made of tiny blocks of ice and dust.

☞ Although just yards thick, Saturn's rings are so wide that if you gathered together all the dust in them you could build a world that is 120 miles across—and astronomers think they may be the remains of a moon that was blown apart by a comet or an asteroid.

☞ Saturn has a very hot interior—up to 22,000 °F—and radiates more energy into space than it receives from the Sun.

Magnificent Moons

Huge lumps of rock whiz round most of the planets, not just ours.

TRUE MOON FACTS

☞ The Earth's Moon is the biggest, brightest thing in the night sky. Yet it has no light of its own. It is just a big cold ball of rock, and only shines because it reflects the light of the Sun.

☞ All over the Moon are large, dark patches that people once thought were seas. So they are all called "mare," from the Latin for sea. Now we know they are vast, dry plains formed by lava from volcanoes that erupted early in the Moon's life.

☞ The Moon's gravity is 17 percent of the Earth's, so astronauts on the Moon weigh the same as a small child on the Earth.

☞ The Moon takes 27.3 days to circle the Earth, but the period between one full moon and the next is 29.53 days, because the Earth moves as well as the Moon. A lunar month is 29.53 days.

☞ The word month comes from the Moon, which takes roughly a month to go around the Earth.

➤➤ **People talk about a rare event as "once in a blue moon." A blue moon is a full moon occurring for the second time in a month. This happens only once about every 33 months.**

➤➤ **Sometimes smoke in the Earth's atmosphere makes the Moon look blue.**

The Earth's Moon has no atmosphere or wind and the dust on its surface never moves. So the footprints left in 1969 by the Apollo astronauts are still there today, perfectly preserved.

More Moons

There are over 150 known moons in the Solar System.

All the planets but Venus and Mercury have moons or "satellites" circling them. Jupiter has 63.

Saturn's moon Iapetus is white on one side and black on the other.

Most of Uranus's moons have girls' names, but they aren't that pretty. Miranda is the ugliest moon of all. It was partly blown apart in the early days of the Solar System.

Mars has two tiny moons: Phobos (pictured) and Deimos. They were named after the horses that pulled Mars' war chariot in Roman mythology.

Stupendous Stars

The stars in the night sky are huge
shining balls, just like our Sun.

> ➤➤ There are dark patches
> on the Sun's surface called
> sunspots. They look dark
> because they are slightly less hot.
>
> ➤➤ Each square inch of the
> Sun's surface burns with the
> brightness of about half a
> million candles!

The night's brightest star Sirius, also called
the Dog Star, has a white dwarf companion
called the Pup Star. Sirius appears so bright
as it is closer to us than other large stars.

TRUE STAR FACTS

☞ Black dwarf stars
are very small, cold,
dead stars.

☞ Red giants are old
stars that change color
as they cool from white to red, and
swell up as their fuel runs out. The biggest stars go
on swelling until they become supergiants that have
swollen to 100 times bigger than they were.

☞ Small stars shine softly for 200 billion years.
Medium-size stars go on for about 10 billion years.
Big stars shine intensely for just 10 million years.

☞ The heart of a star reaches about 29 million °F.
A grain of sand this hot would kill someone
standing 100 miles away.

☞ The biggest stars of all are 100 times as big as
the Sun and 1,000 times as bright.

Under the Sun

☞ The Sun is 93 million miles from the Earth. Because it is further away, it looks the same size in the sky as the Moon, even though it is 400 times bigger.

☞ The temperature at the surface of the Sun is 10,936 °F.

☞ Heat from the Sun's interior erupts on the surface in patches called granules, and gushes up in flamelike tongues called solar prominences, such as the one pictured below.

Just Another Star...

✎ The Sun is a star, just like all the stars in the night sky. In fact, it is a medium-size star in the middle of its 10-billion-year life. But it is close to us, just 93 million miles away, not billions of miles away like other stars.

✎ Like the Earth, the Sun has a core, where most of its heat is generated. The heat takes 10 million years to rise to the surface. This heat turns the surface into a raging inferno, which burns so brightly that it floodlights the Earth and gives us daylight.

✎ The Sun is 100 times as wide as the Earth.

➤ The hypergiant Cygnus OB2 No. 12 is the biggest known star. It's as bright as 810,000 Suns.

➤ The tiniest shining star is still 100 times bigger than Jupiter.

Space Junk

Billions of lumps of rock and ice float around our Solar System.

Meteoroids and Meteorites

☞ Meteoroids are chunks of rock and iron that have come away from larger asteroids and comets.

☞ Many meteoroids collide with Earth, but they are usually so tiny that they burn up as they crash into Earth's atmosphere.

☞ Occasionally, there are chunks large enough to make it all the way down to the ground. These chunks are called meteorites.

☞ Every 50 million years or so, Earth is hit by a rock measuring over 6 miles across.

Asteroid Ida is about as big as New York City and even has its own moon. The craters on its surface show that it is relatively old.

➤ Most meteorites are smaller than a fist and barely noticed. But a few are much, much bigger.

➤ These can wreak havoc when they strike the ground—not only creating giant impact craters, but causing devastation for all kinds of life-forms.

Comets may look amazing, but they are just large, dirty ice balls a few miles across. Nonetheless, they can destroy meteoroids that get in their way, and would cause climate change on the Earth if they scored a direct hit.

Shooting Stars

As meteoroids burn up they leave bright glowing trails in the night sky, earning them the name meteors or shooting stars.

Every now and then, Earth collides with a big clump of meteoroids, creating a shower of shooting stars in the night sky.

TRUE COMET FACTS

☞ For most of their orbit, comets are far from the Sun, but when they swing in close, the ice in them melts, throwing out a tail of dust and gas millions of miles long.

☞ The comet with the longest ever recorded tail is the Great Comet of 1843. Its tail stretched over 500 million miles! This is about the same distance as Earth is from Jupiter.

☞ Comets were named after the Greek word for hair, because of their long tails.

☞ The biggest crash ever witnessed occurred when the comet Shoemaker-Levy 9 smashed into Jupiter in July 1994.

☞ Comet Hyakutake gives off 9 tons of water per second as it passes near the Sun.

☞ Halley's Comet can be seen from Earth every 76 years. Its appearance has often wrongly been thought to be the cause of dramatic events. It last passed by Earth in 1986.

Race to Space

In the 1950s, the USA and the Soviet Union went head-to-head in the Space Race.

With no gravity to pull the ink down, ordinary pens don't work in space. To solve this, the U.S.'s North American Space Agency (NASA) bought special zero-gravity pens for their astronauts to use. In the Soviet Union, they had a simpler solution—their cosmonauts just used a pencil!

➤➤ The first person to orbit Earth was the Russian Yuri Gagarin (left), in April 1961.

➤➤ He survived the space flight but was killed in a plane crash seven years later.

Weightless Wonders

✎ Going to the bathroom is a problem in the weightlessness of space. Poop needs to be sucked away and freeze-dried, then bagged and stored to stop it from floating around inside the spacecraft.

✎ Astronauts don't need a bed in space. They simply float, tethered by a few straps.

Shuttle Services

☞ The rocket boosters on NASA's Space Shuttle each burn 4½ tonnes of fuel per second.

☞ The shuttle's main engine delivers almost 40 times as much power as a train's, but weighs only about one-seventh as much.

☞ When on the launch pad, the shuttle and its boosters tower 190 feet high—39 feet taller than the Statue of Liberty.

☞ To stay in orbit 125 miles up, the shuttle has to fly at nearly 18,650 mph.

☞ It takes the shuttle only about eight minutes to shoot away from Earth and accelerate to this orbital speed.

Animals in Space

✎ The first living creature in space was a dog called Laika (above), who went up in the Russian Sputnik 2 in 1957. Sadly, she could not be brought back to Earth.

✎ Scientists found that when a frog throws up in space, because of the low gravity, it throws up not just its food but its entire stomach. After throwing up, the frog's stomach dangles from its mouth, so it empties out the food and swallows the stomach back in again.

TRUE SPACE SUIT FACTS

☞ Astronauts have to wear a space suit every time they venture outside their craft.

☞ Out in the vacuum of space, an astronaut's blood would boil if it weren't for the space suit, in which the middle layers blow up like a balloon to press hard against the body.

☞ Durable NASA spacesuits cost about $22 million each to make. Now astronauts often wear cheaper $1.8 million suits that wear out after about 461 hours in space.

☞ Astronauts become a little taller in space. There is less gravity, so their bones are less squashed together. They need to take account of this when they're being fitted for their suits—they'll grow into them eventually!

The Enormous Universe

Our galaxy is just one of billions in an ever-expanding Universe.

Big and Getting Bigger

✎ The Universe is the whole of space and everything in it. The Universe contains over 100 billion galaxies (groups of stars), each containing 100 billion stars like the Sun.

✎ The furthest galaxies from the Earth are hurtling away at almost the speed of light.

✎ Scientists now think the Universe began from virtually nothing about 13.7 billion years ago, and has been swelling out at a fantastic rate ever since. This is called the Big Bang theory.

With the naked eye, you can see three other galaxies beyond the Milky Way: Andromeda (above), which is 2.5 light years away,

TRUE MILKY WAY FACTS

☛ The ancient Greeks called our galaxy the Milky Way because it makes a pale band across the night sky, which they thought looks like milk from the breasts of the Greek goddess Hera.

☛ The Milky Way contains 400–500 billion stars. It is 100,000 light years across and 1,000 light years thick, but has a huge bulge 3,000 light years thick at its center, where there are only very old stars and a little dust or gas.

☛ The Milky Way is whirling rapidly, spinning our Sun and all its other stars around at 560,000 miles per hour.

☛ Moving at this speed, our Solar System takes 250,000 years to go around the Milky Way once.

☛ The average distance between stars in the spiral arms of the Milky Way is thought to be 7 light years—that's about 443,000 times the distance between the Earth and Sun.

Spiral galaxies may be spinning round a black hole, which sucks in stars like water spiraling down a drain hole. The galaxies form a fractal pattern like this.

Star Sandwich

The stars in spiral galaxies look like giant fried eggs. But spiral galaxies are actually shaped more like burgers. The rolls are made mostly of invisible "dark matter." The stars are the filling.

➤ Galaxies are often found in groups, called clusters, of up to 1,000.

➤ If you suspend three grains of sand in a sport arena, it will be more closely packed with sand than a galaxy is with stars.

Black Hole

There is a black hole like this one at the heart of most galaxies. A black hole's gravitational pull is so strong that it even pulls light into it.

CRAZY

Nature is full of surprises. For a start, you never know what the weather will bring—from showers of squid to winds that can pluck the feathers from a chicken. Then there are some very strange plants, including squirting cucumbers and plants that eat insects. Discover nature's surprises!

NATURE

Extreme Survivors!

Soggy Rain!

Plants that Eat Meat!

Soggy Rain

We all know that water falls from the sky, but what do you do if it starts raining frogs?

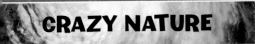

☞ Raindrops are never shaped like teardrops. Usually they are shaped more like balls. The biggest raindrops of all are shaped like doughnuts.

☞ The odors released by fungal spores in the soil give rain its characteristic smell.

☞ Most of the rain that falls around the world falls on the sea. About two-thirds of the rain that falls on land falls in the tropics.

In 1976, Big Thompson Canyon in Colorado was hit by 10 inches of rain in four hours. A flash flood sent 45 million tons of water surging through the canyon.

In Bournemouth, England, in 1948, a group of unfortunate golfers had their game rudely interrupted by a light shower of herring. The fish fell on them out of a completely clear sky.

Weird Rain

✎ In 1873, Missouri was covered with frogs that fell from the sky during a storm.

✎ A Korean fisherman, fishing off the coast of the Falkland Islands, was once knocked unconscious by a single frozen squid that fell from the sky.

➤➤ **The biggest clouds in the world are thunderclouds, or cumulonimbus, which are sometimes more than 12 miles tall and can hold 450,000 tons of water.**

➤➤ **The highest clouds, cirrus clouds, are so high that they are made of ice crystals.**

Downpours

☞ The wettest place in the world is usually Tutunendo in Colombia, which gets 460 inches of rain every year—20 times more than Seattle. Cherrapunji in India can be wetter some years.

☞ The world's rainiest place is Mount Wai'ale'ale in Hawaii, where it rains 360 days a year.

☞ In 1952, 73 inches of rain fell on the island of La Réunion in the Indian Ocean in one day in 1952.

☞ On June 22, 1947, 12 inches of rain fell in 42 minutes on Holt, MO.

☞ In one minute, $1^1/_2$ inches of rain fell on Guadeloupe in the West Indies!

☞ In 1903, a cloud burst sent a 20-foot-high wall of water through the town of Heppner in Oregon.

☞ In February 1998, $21^1/_2$ inches of rain fell on Santa Barbara, California— the wettest California month ever.

In July 2001, a red rain fell on Kerala, India. At first it was thought that a meteor was responsible for the strange-colored rain, but analysis showed that the water was filled with fungal spores.

Gusty Winds

It can get very windy on the Earth sometimes. Don't get blown away!

TRUE WIND FACTS

☛ The Earth is windy because the planet is warmer in some places than others. Hot air rises in warmer places, and cold air is drawn in underneath it. We feel this air movement as wind.

☛ It is nearly always windy at the coast. One reason for this is that land heats and cools much more quickly than water. When the land is warmed during the day, the hot air rises and draws air in off the cooler sea. At night, the land cools and the wind reverses direction.

☛ In Western Australia, the cool wind that blows off the sea is called the Fremantle Doctor because it makes everyone feel better in hot summers.

☛ Seaside golf courses in Britain are specifically designed to be played in windy conditions.

The fastest gust of wind ever recorded was at this weather station on Mt. Washington in New Hampshire on April 12, 1934, when the wind reached 231 mph.

Blowy Places

✎ The average wind speed on George V Island in Antarctica is higher than anywhere else on the Earth. The wind often blows at over 186 mph.

✎ Although Chicago is called the Windy City, the United States' windiest city is actually Dodge City, Kansas.

➤ Tornado Alley in Kansas has more than 1000 tornadoes a year.

➤ England has more tornadoes per square mile each year than any other country, but they are usually small.

➤ Huge sandstorms called haboobs regularly hit towns in Sudan. A haboob dust cloud can be 3000 feet high and 90 miles wide.

➤ In Australia, the bush is often struck by whirling columns of dusty wind called willy willies.

Terrifying Tornadoes

☛ Scientists call a tornado with wind speeds over 261 mph an F5 or "Incredible" tornado.

☛ An F5 tornado can lift a house or carry a bus hundreds of yards.

☛ In 1879, a tornado in Kansas tore up an iron bridge and sucked the river beneath it dry.

☛ In 1955, nine-year-old Sharon Weron and her pony were carried 1000 feet through the air by a tornado. Amazingly they were both unharmed.

☛ In 1990, a tornado in Kansas lifted an 88-coach train from the track and dropped it in piles four coaches high.

☛ Tornadoes can pluck the feathers off chickens.

Horrible Hurricanes

Hurricanes are the most destructive storms of all. They swell up in the tropics and produce frightening wind and rain.

Each hurricane has a name taken from a list issued each year by the World Meteorological Organization. The list is in alphabetical order, so the first storm of the year, for instance, could be Hurricane Andrew. There is one list for Atlantic storms and one for those in the Pacific.

TRUE HURRICANE FACTS

☞ Hurricanes generate the same energy every second as a small atomic bomb.

☞ Hurricanes move from east to west across the ocean to hit east coasts. They unleash torrential rains and winds gusting up to 225 mph.

☞ The average life of a hurricane is nine days. They die out as they move toward the poles into cooler air.

☞ Hurricanes cause huge damage, nearly all of it during their first 12 hours onshore.

☞ At the center of a hurricane is a completely calm area called the eye.

Devastating Storms

☞ In 1970, Bangladesh was struck by the deadliest hurricane ever. The storm caused flooding that claimed over 250,000 lives.

☞ Hurricane Katrina, which struck New Orleans in September 2005, caused over $200 billion of damage in the city. Millions of people were forced to leave their homes.

☞ Much of the damage caused by Hurricane Katrina happened when New Orleans' defensive dykes, the levees, broke and water flooded in.

☞ On October 16, 1987, a hurricane hit southern England, hours after a TV weatherman had said it wouldn't happen.

A storm surge is a rise in sea level caused by hurricanes. In 1899, at Bathurst Bay, Australia, a hurricane produced a 143-foot-high storm surge, the largest ever recorded.

An Angry God

The word "hurricane" comes from the Maya people of Central America who worshipped Huracan, the god of big winds and evil spirits.

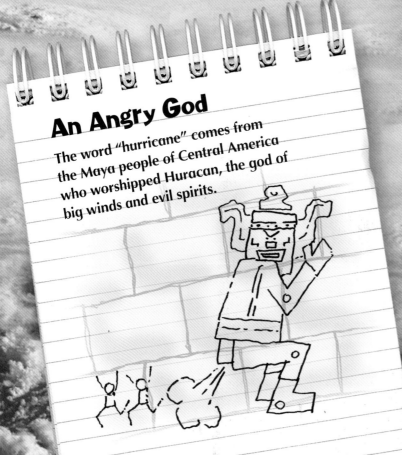

Hurricane Assault

☞ Hurricanes can occur all year round, but in the Atlantic the official hurricane season is between June 1 and November 30.

☞ On average, there are six to eight hurricanes a year.

☞ The record for the most hurricanes in one season is 12, which occurred in 1969.

☞ Hurricanes are also called willywaws, typhoons, and tropical cyclones, but whatever you call them, they're scary storms.

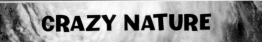

Fatal Flooding

When rivers break their banks, or the sea level rises, it can mean disaster for the people living nearby.

➤➤ Cities can be devastated if the rivers that flow through them flood. The Thames Barrier, pictured, protects London from dangerous high tides.

➤➤ Prague suffered its worst flood ever in 2002 when the Danube River rose by 33 feet.

Asian Floods

☛ Floods around the world can be vital to life. The monsoon floods in India kill around 1,000 people each year, but if the rain doesn't come, millions can starve. In Bengal in 1770, 10 million people died when the rain did not come.

☛ Bangladesh regularly floods because it is on the floodplain of three huge rivers—the Brahmaputra, the Meghna, and the Ganga. Much of the water in these rivers comes from the distant Himalayas, so rainfall far away is crucial to Bangladesh's floods.

Global Warming

✎ After the last Ice Age, sea levels rose 425 feet when the seawater expanded and the ice melted.

✎ Today global warming is causing the water to expand again and sea levels are rising. Whole islands could disappear soon. The people of Tuvalu in the Pacific are already planning where to go.

☞ Tsunamis are giant waves set off by earthquakes under the sea. They can cause huge destruction when they hit land.

☞ A tsunami can race across the ocean at up to 500 mph—faster than a jet plane. In 1960, tsunami waves generated in Chile reached Japan, more than 10,500 miles away, in less than 24 hours, killing hundreds of people.

☞ The tsunami that struck southeast Asia on December 26, 2004 killed nearly 300,000 people.

☞ In 1958, Lituya Bay in Antarctica was hit by a tsunami over 1,640 feet high.

Chinese Floods

➤➤ In 1332, 7 million people were drowned when the Huang He, or Yellow River, burst its banks.

➤➤ In 1887, up to 2 million died when the Huang He flooded.

➤➤ In 1642, 300,000 people drowned when Chinese rebels destroyed dykes near Kaifeng.

➤➤ Over 500,000 Chinese died in 1938 when Nationalist troops broke the banks of the Huang He to stop the advance of Japanese troops.

➤➤ In 1975, almost 250,000 people died when the Banquia Dam on China's River Ru collapsed.

Large areas of the Netherlands, and parts of East Anglia in Great Britain, are below sea level and would be flooded if dykes didn't hold back rivers and the North Sea.

Shocking Lightning

Bright flashes in the sky mean that something in the air is becoming electric.

Lightning Strikes

☞ Lightning is the number one cause of storm-related deaths. In the United States, for instance, you are twice as likely to die from lightning than from a hurricane, tornado, or flood.

☞ If your hair stands up in a storm, get indoors immediately, or inside a car. It may mean the negative charge in the storm is being drawn to the positive charge in you!

☞ Never use a landline telephone during a thunderstorm. If lightning strikes a line outside, the charge will travel up the line and electrocute you through the receiver. Cell phones are OK!

People sometimes say lightning never strikes twice. This is not true. The Empire State Building in New York, above, averages over 20 hits per year. It was once struck by lightning 15 times in the space of 15 minutes!

Claps of Thunder

✎ Thunder is the boom made by air when it is heated up by lightning and expands at supersonic speed. The same effect happens when you burst a paper bag, but it's not quite as loud!

✎ Because sound travels more slowly than light we hear thunder after we see lightning. Thunder is delayed by three seconds for each 1,000 yards we are from the storm.

✎ We can hear thunder from up to 12 miles away.

✎ Thunder often "rolls" because one end of the lightning flash is further away than the other, so its sound takes longer to reach you.

Bolts from the Blue

✏ Very occasionally, lightning emerges from the top of a cloud to strike the ground many miles away from the storm. Long bolts of lightning such as this are called positive giants.

✏ Positive giants are sometimes called "bolts from the blue" because they shoot out of a clear blue sky.

Lightning crawlers, or "spider lightning," can travel over 30 miles as they "crawl" across the bottom of clouds. This is a very rare and very beautiful kind of lightning.

TRUE LIGHTNING FACTS

☛ A flash of lightning is brighter than 100 million 100-watt lightbulbs. For a split second, it has more power than all the power stations in the United States.

☛ Lightning travels at up to 62,000 miles per second—that's 30,000 times faster than a bullet. It moves down a path that is the width of a finger but up to 9 miles long.

☛ The temperature of a typical lightning bolt is four times hotter than the surface of the Sun, reaching over 45,000 °F.

☛ Every day, about 44,000 thunderstorms unleash their ferocity around the world. Lightning strikes somewhere about 100 times every second.

Freezing Cold

As the temperature falls, rain turns to snow, and then it becomes too cold even for snow to fall!

➤➤ More snow falls in the United States than falls at the North Pole because it is usually too cold to snow at the North Pole.

➤➤ The ice that Arctic icebergs are made from is between 3,000 and 6,000 years old.

Bumpy Skin

✎ When you get cold you may suffer from horripilation, or goosebumps. These make hairs stand on end to trap air next to the skin and keep you warm. It works for furry animals, but we have so little hair that it's pretty useless to us.

✎ Wind increases heat loss and makes it feel much, much colder than the thermometer shows. American Paul Siple coined the term wind-chill factor to describe this effect.

Antarctic Chill

☞ The coldest temperature ever measured on Earth was -128 °F at Vostok, Antarctica, on July 21, 1983.

☞ The average temperature at Polus Nedostupnosti in Antarctica is -72 °F.

W. A. Bentley, an American farmer, photographed thousands of snowflakes through a microscope. He found that all snowflakes have six sides, but no two are ever exactly alike.

These girls live in the coldest town in the world—Verkhoyansk in Siberia, where the temperature has been known to plunge to -90 °F.

Fresh Snow

☞ Over 36 million tons of snow can fall in one snowstorm.

☞ In the winter of 1998–1999, 98 feet of snow fell on Mount Baker near Seattle, Washington—a world record for one year. That's enough to bury a large apartment building.

☞ The biggest snowflake ever recorded, over 12 inches across, fell on Fort Keogh, Montana, on January 28, 1887.

Extreme Survivors

Plants can survive in the most extreme places on the Earth, from icy peaks to scorching deserts.

Alpine Plants

☞ Some flowers that grow near the icy peaks of the Alps, such as edelweiss, have woolly hairs, which keep out the cold and protect them from the sun.

☞ Trees just can't survive in the bitter winds of the Arctic—unless they are trees that grow very, very small and close to the ground, such as dwarf willows. They are only a few inches high and are the world's smallest trees.

The flower that blooms nearest the North Pole is the Arctic poppy. In the Arctic, there is so little soil that some plants grow in the bodies of dead animals instead.

➤➤ The giant saguaro cactus of North America grows up to 50 ft and can live for 200 years.

➤➤ One big saguaro cactus may contain 1,320 gallons of water, as much as a small swimming pool.

☞ Whenever it's dry, the ocotillo plant sheds all its leaves so that it does not lose moisture through them. It grows a new set as soon as it rains.

☞ The desert quiver tree goes even further and drops its branches to save water in times of drought.

☞ Some desert plants find water far underground by growing very long roots. The mesquite grows roots as deep as 150 feet.

☞ Pebble plants have found a good way to escape the scorching sun—growing mostly underground!

☞ Window plants go even further underground. They grow downward, not up, leaving just a cool green "window" on top to catch the sun.

Rain Forest lianas need to get high up to reach the sun, but they don't need to grow a big woody trunk—they simply cling on to other trees with little hooks and climb up. A liana can wind its way up to 1,600 feet from tree to tree.

Dead Clever

✎ Most of the time, the leaves of "resurrection plants" look wrinkly brown and dead, but when it rains, they turn green and lush—hence their name.

✎ The Rose of Jericho plant dries into a tight ball which can last for years. As soon as it rains, the ball opens and the plant sprouts.

Volcanic Flower

The silversword is a shrub that lives on the tops of old volcanoes in Hawaii. Only about $2^3/_4$ inches of rain falls each year so the silversword may take 20 years to store up enough water to flower. A few weeks after flowering, the plant dies.

153

Extraordinary Plants

Plants come in an amazing variety of shapes and sizes.

TRUE ANCIENT FACTS

☞ The ginkgo, or maidenhair, is the oldest known plant that grows from seeds. It probably first appeared in China in the Jurassic Period some 180 million years ago.

☞ When scientists planted some 10,000-year-old Arctic lupin seeds found frozen in the soil in the Yukon, Canada, some grew into plants—and one even flowered.

☞ Lichens may be small but they live a long time. They survive in cold places and grow very, very slowly. Some lichens in Antarctica may be around 4,500 years old.

Biggest and Smallest

✐ The world's tallest flowering plant is the Australian Mountain Ash—one was 375 feet tall.

✐ The largest leaves in the world belong to the raffia palm—up to 65 feet long.

✐ The biggest flower head of all is that of the Andean *Puya raimondii* bromeliad. It can be up to 8 feet across and 33 feet tall, and have 8,000 individual blooms. The plant takes 150 years to grow its first flower, then it dies.

✐ The duckweed *Wolffia arrhiza* floats on ponds and is the world's smallest flowering plant. You could fit 25 of these plants across your fingernail. A bouquet of a dozen blooms would fit on a pinhead.

Fruity Fuel

☞ The avocado is the world's most energy-packed fruit, with 60 kilojoules in just $\frac{1}{3}$ ounce.

☞ The world's least energy-rich fruit is the cucumber which gives you just 6 kilojoules per $\frac{1}{3}$ ounce.

Weird Plants

☛ The *Welwitschia mirabilis*, in southern Africa's Namib desert, lives for centuries with just two leaves and no roots, soaking up water through its leaves.

☛ Stems fall off the jumping cholla so easily that people think they actually jump on you.

☛ Two orchids in Australia bloom underground, out of sight of bees. Nobody knows how they pollinate.

☛ In 1982, Soviet cosmonauts managed to get an arabidopsis plant to grow, flower, and produce seeds in zero gravity.

The most ancient of all flowering plants are magnolias. Fossilized magnolias have been found in rocks that are 20 million years old.

Big Lily

The world's largest water plant is the giant lily that grows in the Amazon. Its huge floating leaves grow the size of cartwheels, and you can actually run across them over the water if you're not too heavily built.

Terrific Trees

Trees are the woody plants that give us lovely oxygen and take away nasty carbon dioxide.

Grow Up, Down, and Out

☛ The *Albizia falcataria* is a tropical pea tree and can grow 33 feet in little more than a year.

☛ The wild fig tree in Transvaal, South Africa, grows roots 400 feet long.

☛ The canopy of a single great banyan tree in Calcutta's Indian Botanical Garden spreads over 3 acres.

In California's White Mountains, there is a bristlecone pine tree thought to be 4,700 years old. In Nevada, a bristlecone that was 5,100 years old was cut down in 1964.

Oldies

➤➤ Southwestern Tasmania is home to the world's oldest bush: the 43,000-year-old King's Holly.

➤➤ In 1994, a hiker discovered a wollemi pine growing in the Blue Mountains in Australia. It was only known previously from 120-million-year-old fossil leaves, so this was like discovering a dinosaur alive and well.

➤➤ Conifers first appeared over 300 million years ago—long before dinosaurs—and there are fossils of species of conifer still around today dating from 120 million years ago.

Survivors

✎ Throw axes, fire, storms, and insect attacks at it, and the ombu tree of Argentina, the world's toughest tree, can survive them all.

✎ Yew trees survive for thousands of years because their slow, twisting growth gives their trunks great strength. They can slow their growth to prevent them from growing so big that they would be damaged by storms.

☞ The world gets through enough wood each year to build a woodpile as big as a football stadium and as high as Mt. Everest.

☞ If we keep on cutting down the rain forests at the current rate, half of what's left will be gone by 2020.

☞ It would take about an acre of forest to soak up the carbon dioxide pumped into the air by the average car in a year.

☞ About an acre of trees also adds enough new oxygen to the atmosphere for 18 people to breathe for a year.

☞ When it dies, a tree returns all the carbon dioxide it has absorbed in its lifetime to the atmosphere. Whether the tree burns or rots, exactly the same amount of carbon dioxide is released.

➤➤ A mature deciduous oak tree grows a quarter of a million new leaves every spring—and loses all of them in the autumn.

➤➤ A single oak tree will produce about 50,000 acorns in a good year.

Big Trees

☞ General Sherman is a giant sequoia, a redwood tree, like these, in Sequoia National Park, California. It's the largest living thing on the Earth: 275 feet tall, 80 feet around the trunk, and weighing in at 3,000 tons.

☞ The tallest tree ever known, a eucalyptus in Watts River, Victoria, Australia, was measured in 1872 at over 490 feet.

Plants that Eat Meat

They can't get what they need from elsewhere, so these plants have turned to the animal world for their food.

TRUE BOG FACTS

☞ Soil in wet areas can be poor, so plants eat meat to get the nutrients they need.

☞ Sundew plants catch insects on sticky tentacles on their leaves, which curl over and glue the insects in place.

☞ Butterworts get their name from the drops that ooze out on their leaves, which glisten like butter. Flies then stick to the leaves, which slowly digest them.

Venus fly traps have leaves that snap shut in a fraction of a second to trap their victim inside.

Pitcher Plants

☞ The vase-shaped leaves of pitcher plants ooze sweet nectar to lure in unsuspecting insects. The insects then slip on the wax inside the vase to drown in the water at the bottom—fully prepared to be digested.

☞ Each leaf of a pitcher plant may trap a dozen or more insects—and maybe even a lizard or two—in a single day.

Sneaky Traps

The 3-foot-tall cobra lily resembles a cobra with its hood spread. Protruding "fangs" cover its "mouth," luring prey with nectar. Once bugs fall inside, they drown and are slowly digested.

A bladderwort's animal traps are little pouches or bladders on its leaves and stems. A trapdoor springs open when any tiny animal touches hairs around the entrance.

Any insect sipping the nectar of the yellow trumpet pitcher plant is instantly paralyzed, then tumbles into the pool to be digested.

The Big Pong

The "corpse flower" is not a meat eater. It just smells like rotten meat, which makes it about the worst smelling flower of all. Flies like it, so they come to visit and spread its pollen.

Monkey Cups

The largest meat-eating plant is the nepenthes, a relative of the pitcher plant, which dangles from vines in the rain forests of southeast Asia. The pitchers are big enough to drown a rat and are sometimes called "monkey cups" because monkeys have been known to drink water from them.

159

Funny Fungi

Fungi often grow in soil like plants, but they're an entirely different kind of organism.

TRUE POISON FACTS

☞ Poisonous mushrooms are called toadstools. There are about 75 kinds of toadstool.

☞ The fly agaric toadstool, below, was once used as fly poison.

☞ The most poisonous of all mushrooms is the destroying angel, which usually kills anyone who eats one.

Waste Disposal

✎ Fungi feed off other organisms, or off the waste they produce.

✎ *Pilobolus* fungi live in cow dung. These fungi, called "decomposers," break down the poop by feeding on it.

Athlete's foot is a condition caused by a fungus growing between the toes. Fungi such as *Candida* and *Pityrosporum* can grow on and inside your body, causing skin infections.

Mold

Mold is actually microscopic fungi, and it's mold that gives cheeses such as Roquefort, Stilton, and Danish Blue their streaky look and strong flavor. Some molds, such as penicillin, are a source of important medicines, such as antibiotics.

Mushroom Magic

☛ Often all you can see of a fungus is a lump called the fruiting body, where the spores it uses to reproduce are made.

☛ The bit of a mushroom we eat is just its fruiting body. The threads that it grows from can live underground for many years.

☛ A field mushroom's cap can grow up to 20 inches across.

Puffballs

☛ Puffballs get their name from the way they puff out all their spores from a big round fruiting body when they burst.

☛ An average-size giant puffball can puff out about 7 trillion spores.

Fairy Rings

✎ Rings of bright green grass were once said to have been made by fairies dancing and were called fairy rings. In fact, the grass is turned greener by chemicals released from the threads of toadstool mushrooms as they spread out in a circle underground.

✎ Some fairy rings are many centuries old.

Seedy Plants

Many plants reproduce using seeds. They may need help to spread their seeds far and wide.

Seductive Orchids

☞ Many plants offer insects rewards, such as sweet nectar, so that they will visit their flowers and carry away pollen. Orchids also use a number of tricks to seduce them.

☞ The flying duck orchid gets its name from the ducklike bill on its blooms. The bill lures in insects with its scent, then closes, trapping the insect until it is covered with pollen and is let out.

☞ Wasps attack the "beard" of the beard orchid, right, thinking it is a cuddly female.

➢➤ The seeds of the sycamore tree have wings, which rotate like a helicopter when they fall.

➢➤ If the wings catch a wind, the seeds can travel away from the tree, ensuring that some will land in a place that is good for growing a new tree.

TRUE SEED FACTS

☞ Fruit helps flowers spread their seeds in poop. Birds and bats eat the fruit, and the seeds come out the other end ready to start growing, with the poop around them the ideal fertilizer.

☞ Unlike birds, bats poop while they are flying, which means they spread seeds more widely. In this way, bats help spread forests.

☞ In Amazonia, the rivers often flood the forest. The trees drop their fruit into the water, and fish eat them up, to poop out the seeds elsewhere.

☞ The Amazon's tambaqui fish has such strong jaws that it can crack the nuts that fall into the river.

Big Seed

The biggest seeds belong to a palm tree called the coco de mer. Its seeds can weigh as much as 44 pounds.

Clock Flowers

Some flowers time the opening and closing of their blooms to ensure the flies that will carry off their pollen are around. Opening and closing times:

❀ Dogrose (*Rosa majalis*) 4:00 to 5:00 a.m.—7:00 to 8:00 p.m.

❀ Chicory (*Cichorium intubis*) 4:00 to 5:00 a.m.—2:00 to 3:00 p.m.

❀ Dandelion (*Taraxacum officionale*) 5:00 to 6:00 a.m.—2:00 to 3:00 p.m.

❀ Potatoes (*Solanum tuberosum*) 6:00 to 7:00 a.m.—2:00 to 3:00 p.m.

❀ Flax (*Linum usitatissimun*) 6:00 to 7:00 a.m.—4:00 to 5:00 p.m.

Nutcracker birds bury seed to save for winter when food is scarce. But because they don't come back for all of them, the seeds are left to grow into new plants. A nutcracker may plant 30,000 seeds in a year.

Seed Scatterers

☞ Squirting cucumbers don't rely on any animal to spread their seeds. They just burst and shoot out their seeds at speeds of up to 60 mph.

☞ Some fruits, such as those of geraniums and milkweeds, simply explode, too, showering seeds in all directions.

☞ Both fruits and seeds can have wings to help them spread far and wide. Willow seeds are shaped like tiny parachutes.

☞ Coconuts float and can drift thousands of miles on ocean currents before they are washed up on land.

AWESOME

ANIMALS

Heard about the bird that paints his bedroom blue for his mate? Or the beetle that can lift 840 times its own weight? What about the jellyfish with a poison so toxic it can kill you in 30 seconds? Find out about these and many other amazing, dangerous or just weird beasts.

Interesting Insects!

Creature Comforts!

Animals on the Edge!

The Buzz on Birds

Birds all have feathers and wings, but some struggle to get off the ground, while others don't even try.

Tropical Birds

☞ No bird has a bill bigger for its body than the toucan (right). It is full of holes to keep it light enough to stop the toucan from tipping over. Even so, the bird has to turn its head around and lay its bill down its back to avoid falling over in its sleep.

☞ Hummingbirds can hover and fly up, down, or backward by beating their wings dozens of times a second.

☞ Condors have great noses. While flying over thick forest, they can smell a dead body on the ground.

➤➤ The largest flying bird is the wandering albatross, which flies the oceans around the Antarctic. Its wings stretch out over 10 feet.

➤➤ The flightless elephant bird of Madagascar really was as tall as an elephant, growing up to 11½ feet tall. Sadly, it's now extinct.

Birdbrains

✎ Scientist Irene Pepperberg trained an African gray parrot called Alex to identify about 50 objects and to ask for each of them in English when he wanted it.

✎ There is no better mimic in the world than a lyrebird. It can imitate 12 other birds so well even the birds are fooled. It can pretend to be a photographer's camera, a car engine, or a car alarm. It can even mimic the chainsaw coming to chop down its habitat.

✎ An ostrich's eye is bigger than its brain!

Eggs

➤➤ An ostrich egg's shell is six times thicker than that of a chicken's egg.

➤➤ An adult human could stand on the end of an ostrich egg without it breaking—but it might fall over!

➤➤ The biggest egg ever laid was that of the elephant bird. It weighed as much as 200 chicken's eggs.

➤➤ The eggs of a bee hummingbird, the world's smallest bird, are no bigger than your fingernail.

➤➤ Cuckoos are sneaky. They lay their eggs in the nests of other birds, who bring up the chicks.

Erratic Tinamous

✎ Tinamous, birds that live in Souh America, are the madcaps of the bird world. They are very shy birds. They rarely fly but when they do, they fly very fast to avoid being seen.

✎ Unfortunately, they are such erratic flyers they sometimes crash into trees and kill themselves.

✎ Tinamous get tired easily, so they have to resort to running instead of flying. The problem is that they sometimes do that over water, which causes quite a commotion, because they suddenly find themselves swimming. Even when they run on land, they tend to fall down.

Rapid Roadrunners

☛ Roadrunners are birds that live in the deserts of the southwestern United States. They are not much good at flying, but they can run as fast as an Olympic athlete.

☛ Roadrunners are often seen tearing along roads at great speed. They are so quick, they can catch and kill rattlesnakes, which they then swallow whole.

Cold-blooded Beasts

Just because they have cold blood doesn't mean they're any less weird.

Craggy Crocodiles

☞ The crocodilians—crocodiles, caimans, and alligators—lived alongside the dinosaurs 200 million years ago and are also reptiles.

☞ Since the dinosaurs died out, the saltwater crocodile has been the world's biggest reptile. It grows to around 20 feet long.

☞ To help them stay underwater without tipping over or floating up to the surface, crocodiles often swallow stones.

Crocodiles are often said to cry after eating their victims. In fact only saltwater crocodiles cry, and they do it to get rid of salt, not because they are sorry.

TRUE FROG FACTS

☞ A typical frog can jump about 10 feet from a standing start—the equivalent of you jumping 50 feet without first running.

☞ Don't kiss a poison dart frog—the poison in its skin could kill you!

☞ The world's largest frog is the goliath frog of West Africa, which grows to over 12 inches long. The Australian cane toad (right) is even bigger—almost as big as a rabbit!

Lizards at Large

☛ The chameleon's tongue is the ultimate flycatcher. To catch a fly as it passes by, the chameleon unrolls a tongue that is longer than its body in a split second.

☛ The basilisk lizard's ability to walk on water earns it the nickname the "Jesus Christ lizard."

☛ The world's biggest lizard is the Komodo dragon of Indonesia, which weighs up to 300 pounds. It's so big it can swallow pigs whole.

➤➤ Reptiles were the first large creatures to live entirely on land, over 350 million years ago.

➤➤ The dinosaurs, the largest-ever creatures on land, were reptiles. The brachiosaurus was over 75 feet long and weighed 40 tons.

Slithering Snakes

✎ Snakes kill more people than any other animal.

✎ The African python can swallow a whole adult impala (a kind of antelope), horns and all.

✎ Snakes called constrictors, such as pythons and boas, don't poison their victims. They simply coil themselves around and squeeze until the victim suffocates.

✎ Confusingly, the slowworm isn't a worm. It looks like a snake but isn't a snake either. It's really a legless lizard.

Marvelous Mammals

Humans are far from being the only strange mammals on the Earth.

Monkey Business

☞ Howler monkeys (right) are the world's loudest land animals. They can be heard 3 miles away.

☞ Primates and elephants may be the only animals that recognize themselves in a mirror.

☞ When a male orangutan burps, it's not to be rude. It warns other males to keep away.

Big Cats

☞ Cheetahs are the world's fastest sprinters, reaching 68 mph. But they can keep this speed going for only 10 or 20 seconds.

☞ If you could jump up to a second-floor window, you might be able to match a puma, which can leap over $16\frac{1}{2}$ feet.

Polar bears' favorite food is seal meat. They love it so much that they'll throw a tantrum if a seal gets away, kicking the snow and hurling chunks of ice around.

Vampire bats in tropical South America get their name because they suck blood from animals, such as cattle, at night. If their host is furry, they'll give the fur a trim before they start their meal!

➤➤ **Strangely, the only other animals to have fingerprints similar to humans are koala bears.**

➤➤ **The smallest mammal in the world is the bumblebee bat, which weighs just 1/15 ounce—lighter than a penny coin!**

More Mammals

✏ The blue whale is the largest animal in the world. Its heart weighs 1,540 pounds—that's 10 adult humans!

✏ The tiny hero shrew really is a hero. Its backbone is so strong it can stand being trodden on by a human!

✏ A male lion can drag a 660-pound zebra. It would take six people to do this.

✏ Rabbits breed like, well, rabbits. A doe rabbit can have 20 babies a month and her babies will have babies after six months. If all the babies survived to breed, a single rabbit could have more than 33 million offspring in just three years!

Elephants

☞ There is no bigger land animal than an African elephant. Big bulls can weigh 13,000 pounds—about as much as eight family cars.

☞ Elephants have bigger brains than any animal other than a whale—much bigger than a human's!

☞ Elephants grieve when one of their group dies.

☞ Elephants have no problems crossing deep rivers—they stick up their trunks and use them like snorkels.

☞ Elephants cannot jump.

Interesting Insects

Take a look at the creepiest, crawliest creatures around!

Rock-hard Roaches

☞ The cockroach's resistance to radiation means it would probably survive a nuclear war.

☞ A cockroach can live for up to a month after losing its head. It will eventually die of thirst.

☞ The ancient Greeks made medicine from cockroaches, grinding them up into a paste. They used the paste to treat earache and wounds.

Termites

✎ Some people claim that termites eat wood twice as quickly when listening to heavy-metal music.

✎ A single colony of African termites can reduce an entire house to rubble in just three months.

✎ Termites build huge mounds to live in that have their own air-conditioning system. The tallest termite mound ever found was like a column—just 10 feet across but almost 45 feet tall.

✎ Termite society is highly organized, and each individual knows its place. There is one king and one queen, whose sole job is to reproduce. The rest of the colony are either workers or soldiers.

✎ A termite queen can live for up to 50 years and lays 2,000 eggs a day.

➤➤ A single army ant can carry 25 times its own weight. Army ants work in groups to carry heavy food back to the nest.

➤➤ Slave-making ants raid the nests of other ants. They steal the young and raise them as slaves.

Honey, which is made by bees to eat and feed to their young, is a naturally made food that does not spoil. Scientists have tasted honey found in the tombs of Egyptian pharaohs—and survived to tell us about it!

TRUE BEETLE FACTS

☞ Dung beetles (right) are the world's sewage operatives. Animal dung never lies around for long before dung beetles roll it away to lay their eggs on. Their young will then feed on the dung when they hatch.

☞ A pile of fresh elephant dung may contain as many as 7,000 beetles!

☞ The leaf-eating beetle uses a film of oil on its feet to stick itself to leaves. Otherwise it would be blown away when the leaf shook in the wind.

☞ There are 250,000 different species of beetle, including some of the largest and smallest of all the insects. All beetles have a pair of hardened covers that protect their wings underneath.

☞ The heaviest flying insect of all is the goliath beetle of Africa. It weighs as much as an orange and grows up to 5 inches long.

Monarch butterflies fly 2,500 miles from North America to find their grandparents' exact birthplace in Mexico, although they've never been there before.

Living in the Sea

All kinds of fish and other strange creatures live in the ocean deeps.

Sharks

✎ The whale shark is the biggest fish in the sea. Some whale sharks may be 43 feet long and weigh over 20 tons. But they are completely harmless, feeding mainly on plankton.

✎ The largest meat-eating fish is the great white shark (left). Great whites almost 23 feet long have been caught, but they may grow up to 30 feet.

✎ A great white shark can bite with a force of about 20 tons per square inch—that's enough to easily bite through steel plate.

✎ Sharks have a better sense of smell than any other kind of fish. They can detect one part of animal blood in 100 million parts of water.

Molluscs

☞ The blue-ringed octopus may be no bigger than a golf ball, but its poison can kill a person in a few minutes.

☞ People once thought the huge giant squid might be a myth, but one was filmed alive for the first time in 2006.

☞ The aptly named colossal squid is even bigger. In February 2007, one measuring over 30 feet was caught.

Giant clams can really live up to their name. One found on the Great Barrier Reef near Australia was over 3 feet across and weighed more than 560 pounds.

Marine Mammals

☞ There is no bigger animal on the earth than a female blue whale. At over 100 feet long and weighing around 160 tons, she is 25 times heavier than the largest elephant.

☞ Blue whales are 26 feet long when they are born.

☞ The blue whale lives on tiny animals just an inch or so long—krill—but it eats up to 4 tons of them every day.

☞ Dolphins have been known to save drowning swimmers by lifting them to the surface, probably because dolphins have a natural instinct to push injured dolphins to the surface so they can breathe.

➤➤ Coral is actually an animal. A coral reef is a colony fixed to the seabed. Algae living on the coral give the reefs their vivid colors.

➤➤ Many kinds of coral feed on phytoplankton—microscopic organisms in the sea, which are food to a lot of creatures, big and small.

Creature Comforts

Animals can make their homes in the oddest of places.

☞ The greater tree swift has one of the smallest nests in the world, made from strips of bark, which it glues together into a small cup. The bird then sticks the nest to a high branch.

☞ The hooded parrot and the golden-shouldered parrot of Australia make their nests by burrowing into termite mounds.

☞ Many birds are nocturnal, but South America's oilbird takes it to extremes. It lives in caves by day and comes out only at night to find fruit from trees, using its great sense of smell.

Young pearl fish live inside sea cucumbers. When a cucumber opens its breathing hole, the pearlfish nips in tail first—then eats the cucumber from the inside.

Big Towns on the Prairie

☞ Prairie dogs live in huge burrows called towns. With hundreds of millions of inhabitants, they're more like megacities!

☞ Within each town, there are nurseries, bedrooms, bathrooms, and a guard room. Prairie dogs live in families called coteries— one male, many females, and all their cubs.

Home Truths

☞ Lizardlike reptiles called tuataras often shack up with seabirds called petrels in their burrows. It's a neat arrangement. The tuatara sleeps by day, and emerges at night to feed. The petrel hunts out at sea by day and sleeps in the burrow at night. It's not always harmonious, though— the tuatara sometimes eats the petrel's eggs.

☞ Thousands of feet down on the ocean floor, there are vast colonies of huge clams feeding off the chemicals that steam up from hot volcanic vents called black smokers. Tube worms can grow up to 6½ feet long around the vents.

➤➤ Leaf-cutter bees live alone, but most bees and wasps, like honeybees and bumblebees, live together in vast colonies.

➤➤ Paper wasps build huge nests, such as this one, out of a kind of papier-mâché, which they make by chewing up wood.

Nests To Let

Social weaverbirds live together in huge treetop nests, where each pair of birds has its own private chamber and entrance. Sometimes other kinds of bird, such as lovebirds, pygmy falcons, and red-headed finches, may move in to share the nest. The weaverbirds don't seem to mind too much, and don't even charge rent!

No mammals (apart from humans) build more elaborate homes than beavers, with their lodges protected by a dam. The dams can be 1,000 feet long and may be centuries old. The room inside the lodge is about 2 feet long and 6½ feet high and wide.

Animals on the Edge

Animals can get by in the most extreme conditions on the planet.

In Japan's icy winters, macaque monkeys keep warm by having a bath in hot volcanic springs. They learned to do this by copying humans—they also make snowballs like us and throw them at each other!

Antelope jack rabbits keep cool in the hot deserts of the southwestern United States by circulating blood through their giant ears, which helps them to lose heat.

Surviving Drought

☛ The addax, an antelope, never drinks. It survives in the Sahara by getting all its water from its food.

☛ The Mojave squirrel of the United States survives drought by going to sleep for days on end.

☛ Kangaroo rats in California's Death Valley save water by eating their own droppings—yum!

☛ Camels can go up to two weeks without drinking, but when there's water around they can gulp down 45 gallons in a day.

Snowed Under

☞ Snowshoe hares are well-equipped for snow. They have great big hind feet that act like snowshoes and stop them from sinking into the snow.

☞ Reindeer need to eat 26 pounds of lichen a day to survive, but they have a remarkable ability to find it under thick snow.

☞ Polar bears have such thick coats that sometimes they have to roll around in the snow to cool down.

Intrepid Insects

✎ The larvae, or young, of the chironomid can dry out completely, losing their water, for up to 17 years and make a complete recovery when conditions improve. Not surprisingly, these tough flies can live all around the world.

✎ Springtails can live in temperatures as low as -36 °F in Antarctica because their body fluids contain substances that do not freeze easily. However, below 14 °F they cannot move.

✎ Arctic beetles can survive the polar chill even when the temperature drops below -76 °F. Alaskan flies can stand being frozen at similarly chilly temperatures and still survive.

Hot and Cold Lizards

✎ Most reptiles rely on warm sunshine to keep them warm and give them the energy to move. But New Zealand's tuatara, right, copes with the cold where it lives by doing everything very, very slowly. It can go up to an hour without breathing. It also takes 20 years to reach a length of 2 feet.

✎ In the baking dunes of Africa's Namib Desert, the fringe-toed lizard avoids scorching its feet by dancing. Sometimes it rests briefly on its stomach to give its feet a rest.

✎ The marine iguana spends its mornings sunbathing on the rocky shores of the Galápagos Islands. It then dives into the cold water to feed. It quickly loses heat in the water, and by mid-afternoon it's time for another session in the sun.

Natural Born Killers

Be afraid! These are some of the deadliest, most terrifying, and downright nasty creatures on the Earth.

Fearsome Fish

☞ Sharks are scary, but they only kill about 10 people each year. Bees, wasps, snakes, and even dogs kill far more.

☞ Black torpedo rays stun their prey with an electric shock as powerful as the current from a household outlet.

☞ The reef stonefish is the world's most venomous fish. Thirteen spines along its dorsal (back) fin, with venom sacs at the base, can inject a venom that causes agonizing pain. Stonefish are camouflaged to blend in with the seabed, and they are easy to step on.

☞ The box jellyfish's venom could kill you in 30 seconds.

☞ The moray eel, left, hides in crevices ready to pounce on unsuspecting fish. It uses its powerful jaws to tear open its prey, and can cause serious damage to humans with just one bite.

Hysterical Hyenas

Hyenas work in packs of up to 90. They can run down a large animal, such as a wildebeest, and eat the whole animal within 15 minutes, even the bones. When excited, hyenas make a mad cackle noise.

Sneaky Snakes

✎ This African egg-eating snake lives in trees, looking for eggs to steal from birds' nests. It can swallow an egg much wider than itself by stretching its jaws. It has 30 "teeth" in its throat, which break down the shell as the egg passes down the throat.

✎ The largest poisonous snake is the king cobra of southeast Asia, which grows to over 16 feet.

✎ Cobras kill 7,000 Indians a year.

✎ The fer-de-lance snake has 60–80 babies, all of them deadly poisonous from birth.

✎ You don't have to be a killer to scare off your enemies, you just have to look like one. The harmless red and black milk snake looks just like the deadly coral snake.

TRUE INSECT FACTS

☛ When it is threatened, the bombardier beetle spurts jets of boiling chemicals from its abdomen at its enemy.

☛ A swallowtail caterpillar deals with its enemies by bashing them with a stinky forked gland it whips out from a pocket behind its head!

☛ Large digger wasps sting bird-eating spiders and paralyze them, to give a fresh meal to their young.

☛ Honeybees have barbed stingers. These get stuck in the skin when they sting us, ripping open the bee's abdomen so that it bleeds to death. When they fight other bees, though, they can use their stingers again and again.

➤➤ Despite their reputation, most tarantulas are not poisonous. They crush their small victims with their powerful jaws.

➤➤ However, bird-eating spiders, such as this one, are big AND poisonous. The biggest ones are as big as a dinner plate!

Survival Tactics

Faced with dangerous predators every day, some animals have come up with some very clever ways to stay alive.

Absolute Stinkers

☛ The skunk has got smelly warfare off to a fine art, squirting out a foul-smelling liquid from glands under its tail. It can score a direct hit at up to 10 feet, and the liquid can be smelled up to 1,760 feet away.

☛ The tiny stinkpot turtle is no bigger than a saucer, but the smell it gives out to keep predators away would make an elephant reel.

☛ When threatened, the hognose snake pretends to be dead, even giving off a smell like it's dead!

Marine Escapologists

✎ When attacked, the apparently small porcupine fish swallows loads of water and swells up like a spiky basketball. If the attacker isn't frightened out of its wits, it finds the porcupine fish an impossible mouthful.

✎ When in danger, the shrimp fish hides among a sea urchin's stinging spines and pretends to be one of them.

✎ The sea cucumber shoots out its sticky guts to smother predators.

No one's going to kill you if you're already dead, so, when threatened, the American opossum rolls over and lies still with its mouth open and its eyes glazed over.

To scare off attackers, the Australian frill-necked lizard can make itself look three or four times bigger and scarier by blowing up the ruff around its neck.

☛ If a predator grabs one of the arms of a starfish called the brittle star, the arm simply drops off, allowing the brittle star to escape while the predator is distracted. It can lose all five arms and regrow them all again.

☛ The glass lizard has no legs, and if it is attacked even its tail drops off. The tail lies there wriggling like a snake, distracting the attacker while the lizard escapes.

☛ Like the glass lizard, a wood mouse can shed its tail to help it escape, but the tail never grows back again.

☛ The earthworm can replace any of its body segments if it loses them, even its head. So if you cut one in half, it could grow into two new worms.

➤ Chameleons change color when they are angry or frightened, too cold or too hot, or when they are sick.

➤ Humans go pale when they're scared or angry. Octopuses turn white—and then all kinds of other colors to frighten off their attacker.

Masters of Disguise

☛ In the shimmering heat haze of the African plains, a zebra's stripes blur its outline and make it hard to see. However, a leopard's spots have the same effect, and help them sneak up on unsuspecting zebras!

☛ Many caterpillars disguise themselves to look just like the plants they eat to avoid being eaten themselves by birds. Some disguise themselves as twigs, others as leaves.

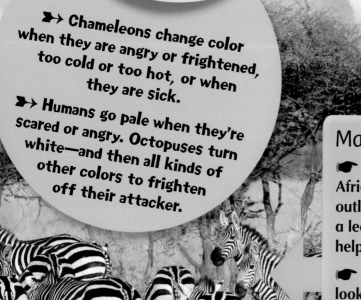

Wild Wooing

Like us, animals can go to extreme lengths to attract a mate and have babies!

Warbling Whales

Male humpback whales are the true romantics of the oceans. They sing elaborate songs lasting 20 minutes or more. Scientists think they are doing this to serenade the females.

The male frigate bird blows out his throat like a big, shiny red balloon to impress the ladies. If she likes it, the female will rest her head on his love cushion.

TRUE WOOING FACTS

☞ When they're in the mood for love, male mice burst into song. Their songs are squeaks too high-pitched to be heard by humans, but female mice are charmed.

☞ Finding your mate in the deepest, darkest ocean can be tricky. So when a male deep-sea angler fish finds his female, he hangs on for dear life and even lives off her blood supply. When she lays eggs, he's on hand to fertilize them.

☞ An Australian bowerbird male builds a love nest called a bower to seduce a mate, decorating it with shells and bones. He paints it a pretty blue with juice from berries, using a piece of bark as a brush.

Mother Love

✎ A mother ichneumon wasp makes sure her young have both a home and food by drilling with her ovipositor (egg-laying tail) into wood to lay her eggs right next to horntail larvae. Once they have hatched, the young can feed on the larvae.

✎ Shrew mothers take their young with them when they go out foraging—the young follow in single file, each holding tightly to the one in front of it.

Doting Fathers

☞ A male Darwin frog (right) swallows the female's eggs and keeps them in his throat until they hatch and pop out of his mouth.

☞ A male midwife toad also looks after the eggs. He strings them together and wraps them around his legs until they hatch.

☞ Giant water bug mothers lay eggs on the father's back and glue them down, so he doesn't lose them.

Demanding Females

☞ In some species of praying mantis, the female often eats the male while they are mating, starting with his head and finishing with his abdomen. By becoming a nourishing meal, the father provides food for the eggs that are his children.

☞ Female bella moths like their males to be dangerous. Males compete with each other and the most poisonous wins. The scents they give off include the poisons from the seeds they ate as caterpillars.

☞ A female North American red-backed salamander tests the quality of a male's lifestyle by dipping her nose in his poop. If he's been eating low-grade ants instead of luxury termites, she can tell and will reject him.

Super Senses

Animals can have amazing senses. Some can even feel the Earth's magnetism.

Eagle-eyed

☛ An eagle can see a rabbit moving on the ground from 1³/₄ miles away.

☛ An osprey can see a salmon moving under water from 100 feet up.

Touchy-feely

✎ Nearly all life-forms respond to being touched, so touch is considered to be the most basic sense.

✎ Mouse whiskers are so sensitive that mice can find their way around in complete darkness using them.

✎ A cockroach's feelers can detect a movement as small as 2,000 times the diameter of a hydrogen atom. We'd need a powerful microscope to see such a small movement.

✎ Honeybees perform dances in the darkness of the hive. Other bees follow every move by feeling with their antennae.

TRUE HEARING FACTS

☛ Crickets can hear with their front legs. Sound waves vibrate a thin membrane on the legs in the same way that they vibrate the eardrum inside your ears.

☛ The noctuid moth can hear sounds with a frequency as high as 240,000 hertz. That's 10 times higher than anything humans can hear.

☛ Elephants can hear sounds as low as 1 hertz, so low humans would feel the vibrations of the sound waves but not hear them.

Super Sniffers

☞ The male emperor moth can detect the smell of a female that is ready to mate at 7 miles, even though the female has only 0.1 micrograms of the attractive aroma, called a pheromone.

☞ A polar bear can smell a dead seal 12 miles away.

☞ A freshwater eel smells its way to its spawning grounds in the Sargasso Sea, detecting key chemicals in the water as dilute as one part in every 3 million.

Echolocation

✎ Bats find their way around in the dark using echolocation. This means that they make short, high-frequency bursts of sound and pick up the way they bounce back off objects to build up a sound picture of their surroundings. It's so accurate that bats can avoid a wire just 1/300 inch thick.

✎ Using their echolocation system, bats can catch two flies a second.

✎ Swiftlets are birds that live in caves and use echolocation like bats, but they make their bursts in clicks at a frequency low enough to be heard by humans.

✎ Freshwater dolphins find their way around in muddy rivers using echolocation.

Bees have magnetic iron oxide crystals in their abdomens which allow them to detect slight changes in the Earth's magnetic field and use them to navigate.

Armadillos

Armadillos use their amazing sense of smell to sniff out termites that are up to 30 inches under the ground.

Animal Olympics

These would be the winners in the toughest events in the natural world's Olympics.

Speed Merchants

☞ The world's fastest animal is the peregrine falcon, which can dive at an incredible 240 mph.

☞ Everyone knows the cheetah is the fastest land animal—but North America's pronghorn antelope runs it a close second with a top speed of over 62 mph. The pronghorn can keep up high speeds much longer than the cheetah, which becomes exhausted after a few seconds.

☞ The cosmopolitan sailfish is the fastest fish in the sea. It can swim 330 feet in just over three seconds.

Highs and Lows

✎ Fish are the deepest swimmers. Some, like the angler fish and the fangtooth, swim around in total darkness at least 16,400 feet down.

✎ The emperor penguin can dive deeper than any other bird. It can plunge to depths of 1,850 feet.

✎ Ruppell's vultures can soar over 36,000 feet. One collided with an aircraft at 37,000 feet.

✎ The bar-headed goose can fly at heights of over 29,500 feet. That's just as well because it migrates over the Himalayas, the highest of which rise to over 26,000 feet.

The sperm whale, left, is a champion diver. It can dive down at least 8,200 feet, and hold its breath for almost two hours. The bottlenose is the champion in the dolphin class. It can dive at least 1,000 feet.

☞ The very toughest animals have to be male emperor penguins, which survive the Antarctic winter in temperatures of -40 °F.

☞ They huddle together to share body warmth, protecting an egg between their feet.

☞ After two months out in the open, they are finally relieved of their egg-guarding duties by the females. All that's left is a 90-mile trek to the sea for their first meal!

TRUE HARD FACTS

Long-haul Flights

☞ The Arctic tern flies 30,000 miles a year as it migrates between the North and South Poles.

☞ The tern takes breaks, though. The Pacific golden plover flies nonstop for 100 hours from Siberia to the South Pacific islands.

☞ An albatross may fly a 3,000-mile round trip to find food for its chick.

Weight Lifters

➤➤ Ants can carry prey over seven times their own body weight.

➤➤ An elephant can pick up a car with its trunk.

➤➤ But the gold medal winners are rhinoceros beetles, which can carry 850 times their own weight. That would be like you lifting up a military tank!

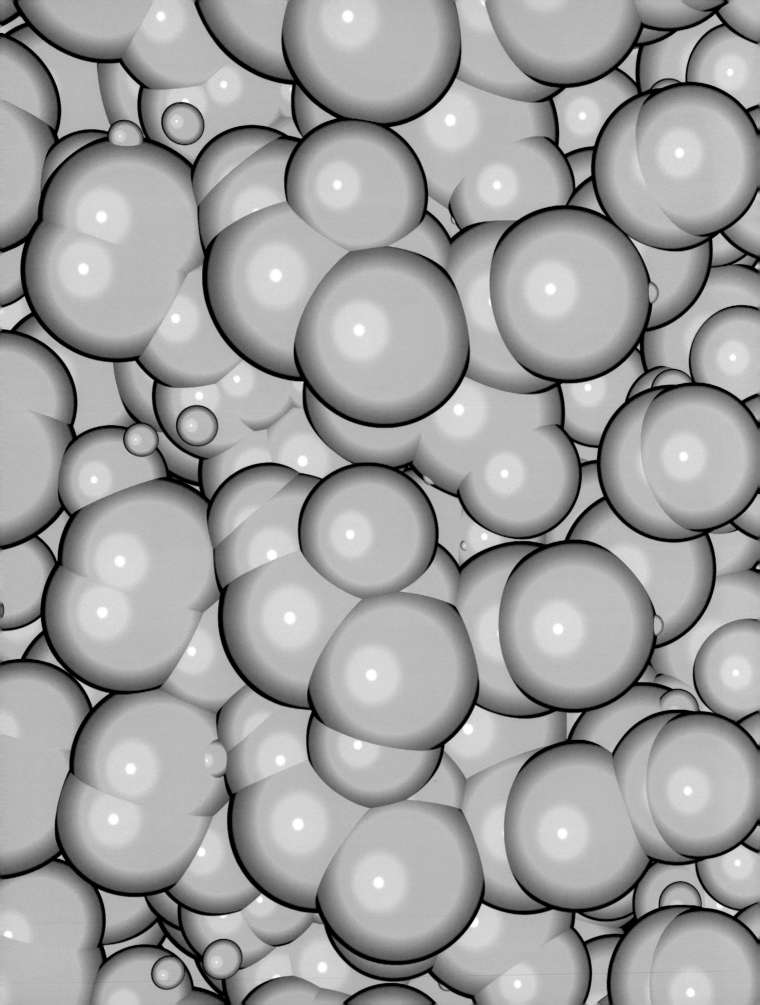

WEIRD
SCIENCE

Science has found some really amazing discoveries. Did you know, for instance, that astronauts age more slowly when they zoom through space, or that you can make spinach taste like chocolate? And as for the scientists—they're a strange bunch, as you'll see in this chapter.

Lovely Light!

Amazing Atoms!

Suspicious Substances!

Feel the Force

Gravity is the force that keeps us all down to Earth!

Gravity in Space

☛ If it weren't for the pull of gravity, the Moon would go spinning off into space away from Earth.

☛ Astronauts in a spaceship in orbit seem to be weightless—but not because they have escaped Earth's gravity. Earth's gravity pulls them very strongly to keep them in orbit. They are falling, but in a way that misses Earth.

☛ Spiders in space spin crooked webs. They need gravity to get the right shape.

Without gravity, if you bounced up on a trampoline, you'd never come back down. You'd just go on shooting up and up and up— and not stop until you bumped into a star.

The Moon's small size and lightness make its gravity weaker than that of Earth. Without their heavy space suits, astronauts would be able to jump up 12 feet on the Moon!

The planet Mars has two tiny moons, whose gravity is so weak you could take off from their surface into space by cycling fast up a ramp.

Universal Gravity

☞ **Gravity is the force of attraction between every piece of matter in the Universe. The more matter something contains, the stronger it attracts other objects with its gravity.**

☞ **Gravity exists between all objects with mass, however large or small they are.**

TRUE FALLING FACTS

☞ **If you fell out of a plane, you'd fall faster and faster for 15 seconds, then no faster. That's because you're traveling at terminal velocity, the fastest you can fall. Hit the ground at terminal velocity and you're terminated!**

☞ **On your way down, you hit air molecules, which slow you down. This is what stops you from falling faster than terminal velocity.**

☞ **Feathers fall slowly because they are shaped to catch a lot of air molecules. In 1604, Galileo predicted that, with no air, a feather and a stone would fall at the same speed. David Scott, on the 1971 Apollo 15 mission, proved Galileo right on the airless Moon.**

Skydivers slow their fall a little by spreading out their body and arms to hit as much air as possible. They do this in free fall, before opening their parachute to catch a lot more air and slow right down.

Exuberant Energy

Energy exists in a lot of different forms. We keep ourselves going by unlocking the energy stored in our food.

Your muscles turn some of the energy in your food into heat energy. This is why your body gives off the heat of 10 light bulbs when you go running.

TRUE OIL FACTS

☞ All our major fuels—oil, natural gas, and coal—obtained their energy from sunlight, absorbed long ago by the plants and marine organisms from which the fuels were formed.

☞ Over millions of years, as these organisms were buried and turned to fuel, the energy became highly concentrated in the form of carbon.

☞ There are 100,000 million million million joules of energy in the world's known oil reserves. That's 1 followed by 23 zeroes.

There is enough energy in a standard barrel of oil to boil 6,600 gallons of water. That's a really big cup of coffee!

Universal Energy

The amount of energy in the Universe, measured in joules, is 4000 million million million million million million million million million million. That's 4 followed by 69 zeroes. Makes you tired just thinking about all that energy.

☞ Energy is measured in joules. One joule gives enough energy to lift an orange by 3 feet.

☞ Energy is never created or destroyed. It only changes from one form to another.

☞ The amount of energy in the Universe that is in the form of heat is increasing, however. So the whole Universe is warming up, not just our planet Earth.

Have you ever seen manure steaming? This happens because there are millions of microbes inside it. The microbes turn its energy into heat, which boils out moisture.

A large egg contains about 400,000 joules of energy. But when people talk about the energy in food, they like to talk in thousands of joules, or kilojoules. An egg contains about 400 kilojoules.

Food Fuel

➤➤ Our energy comes from food, especially sugary and starchy foods, such as pasta, which contain a large proportion of energy-rich chemicals called carbohydrates.

➤➤ Your body uses energy all the time—even when you're sitting still. You get through almost an egg's worth of energy just watching television for an hour.

➤➤ Seven hours of hard physical work uses about 10 eggs' worth of energy.

Lovely Light

Without it our Universe would be a very dark place.

TRUE LIGHT SPEED FACTS

☛ Light is the fastest thing in the Universe. Light travels 983,571,056 feet every second. Light can get all the way from you to the Moon in the time it takes you to say, "Woahh! Hold on!"

☛ A beam of light can get from Earth to Mars in three minutes and to the Sun in eight and a half. It takes just over four years to get to the nearest star, Proxima Centauri.

Light travels more slowly through water than through air. This means that light bends when it passes from air into water, which is why straws appear to be broken.

Photons, or light particles, from the very furthest galaxies are very old indeed. They've been whizzing along for over 12 billion years!

Light Waves

☛ Light travels through the Universe as tiny, invisible waves of energy. You can see light only when it hits something, which is why empty space is dark.

☛ Light waves are so short that 14,000 of them would fit across your thumbnail.

☛ Light waves are formed by countless tiny particles called photons, zig-zagging 600 trillion times a second.

☛ On a sunny day, in just one second 1,000 billion photons of light strike a pinhead.

Fluorescent Lights

☞ Fluorescent lights stay cool and produce a brilliant white light, but they do flicker slightly. This means that the amount of light they give off isn't constant, which can give people headaches.

☞ They produce less heat than normal bulbs, so they use less electricity.

➣ White light, like sunlight, contains every color there is.

➣ When a beam of white light shines through a wedge of glass called a prism, it emerges split into a rainbow of colors called a spectrum.

Amazing Atoms

We used to think nothing was smaller than an atom. Then we found even smaller things. And then smaller still...

Structure of the Atom

☞ Scientists once thought atoms were as small as you could get. Then they found that an atom is made of other particles—electrons, protons, and neutrons.

☞ At the center of the atom is a tiny dense core, called a nucleus, filled with protons and neutrons. Around it whiz just a few electrons, which are even tinier.

An atom is mostly empty space. If the nucleus were a football at the center of a stadium, the electrons would be like peas flying around the rows of seats.

Mighty Atoms

☞ Despite being mainly empty space, most atoms are just about indestructible.

☞ All but the very smallest atoms were forged billions of years ago inside stars.

☞ The atoms in your body were all made in stars and spread through the Universe when the stars exploded. So we are all made of stardust.

☞ Very big atoms, such as uranium, slowly lose parts and change into other atoms in a process called radioactive decay.

☞ Scientists have recently found hundreds more particles that are smaller than atoms. And they are discovering smaller particles all the time.

☞ The smallest particles of all may be ones called quarks. Protons and neutrons are made of quarks. If protons were the size of grapes, quarks would be no more than a hair's breadth across.

☞ Quarks have been given very odd names by the scientists who discovered them. One is called Strange. Another Charm. Another Truth. Another Beauty.

Tiny Things

☞ You could fit 2 billion atoms on the period at the end of this sentence.

☞ There are about 450 trillion trillion atoms in your head.

☞ The number of atoms in the Universe is about 1 followed by 80 zeros.

☞ Atoms combine with each other into larger molecules to make different substances. For instance, two hydrogen atoms combine with one oxygen atom to make water.

More Parts

✎ There are particles that transmit forces of attraction between particles, like love letters. These particles are called bosons.

✎ Some particles effectively glue the quarks in protons and neutrons together. Scientists weren't in a very creative mood when these were named. They are called gluons.

Suspicious Substances

Some chemicals and materials do very strange things indeed.

TRUE SUBSTANCE FACTS

☞ Nearly all substances increase in size as they get hotter. But zirconium tungstate shrinks—scientists still don't know why!

☞ "Mood rings" change color as the temperature changes because they contain thermotropic substances. These are crystals that respond to temperature changes by twisting. As they twist, they reflect or absorb different colors of light.

☞ When it is super-cooled below -458 °F, helium gas not only turns into a liquid, it actually starts to glide up the sides of the container it's in.

☞ Custard briefly acts like a solid if you hit it. A substance like this is described as a non-Newtonian fluid.

Smelly Stuff

✎ In World War II, French Resistance fighters doused German soldiers with bombs filled with a stinky mixture, which they called "Who-me?" (as in, "Someone around here stinks. Who is it? Oh, it's me!").

✎ One of the smelliest natural molecules is butyl mercaptan—the stink produced by skunks to scare off attackers.

✎ To test air fresheners, U.S. chemists developed a chemical that smelled like poop, which they called "Government Standard Bathroom Malodor."

➤➤ Quicklime is a substance often used to treat sewage because it helps make all the sloppy poop dry out.

➤➤ Quicklime is called "quick," Old English for "living," because it twists and swells as if alive when water drips on it.

Glow-in-the-Dark

☞ Fireflies, along with some bacteria and plankton, glow in the dark because their bodies contain a natural chemical called luciferin. The luciferin gives out energy in the form of light when it combines with oxygen in their cells.

☞ In the mid-1800s, people used to take pills containing phosphorus to make them more clever. But the only way it made them brighter was that they glowed a little in the dark!

☞ Modern paper is treated with fluorescent compounds to make it look whiter. So if supposed historical documents give off a fluorescent glow under ultraviolet light, they are probably modern forgeries.

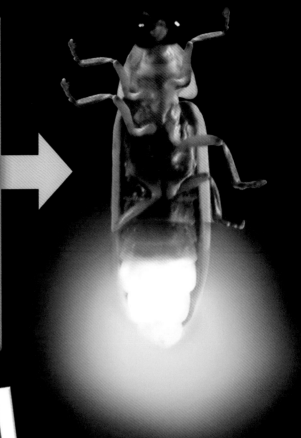

The quinine in tonic water is fluorescent. This means that any drink made with tonic will actually glow a faint blue color in the dark.

It's Called What?

☞ Moronic acid is a kind of resin, a substance made by plants, often found by archaeologists inside ancient Egyptian funeral jars.

☞ Traumatic acid is an acid that occurs in plants and helps plant wounds heal.

☞ Munchnones are compounds with ring-shaped molecules, named after Munich in Germany.

☞ Constipatic acid is an acid that occurs in some lichens in Australia.

Science on the Edge

Even the cleverest thinkers still struggle to understand science's most mind-boggling theories.

In addition to the three dimensions we can see and a fourth, time, that we can experience, the Universe may be made of six or seven more dimensions that we cannot sense at all.

Einstein for Astronauts

Relativity effects are real, not just theoretical. When astronauts went to the Moon, the clock in their spacecraft actually lost a few seconds during the journey. The clock wasn't faulty, they were just traveling fast enough for time to run slower. So the astronauts returned to Earth slightly younger than if they had stayed on the ground!

TRUE RELATIVE FACTS

☛ According to Einstein's Theory of Special Relativity, all speeds are relative, except for the speed of light, which is always the same, no matter how it is measured.

☛ Einstein introduced the idea of "space-time" because his theory of relativity meant you could never analyze movement in terms of distance alone. You had to take time into account, too.

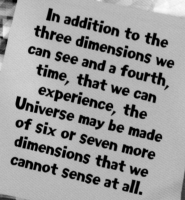

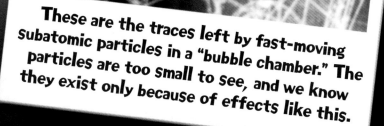

These are the traces left by fast-moving subatomic particles in a "bubble chamber." The particles are too small to see, and we know they exist only because of effects like this.

It's All Relative

☞ If spaceships could ever travel fast enough, Einstein's theories show they would distort space-time as they approached the speed of light.

☞ The distortion of space-time as spaceships traveled very fast means they would actually appear to get shorter and shorter. If they ever got to the speed of light, which Einstein says they can't, they would shrink to nothing.

☞ The clocks on a spaceship would run slower as time stretched (time dilation). If the spaceship ever got to the speed of light, time would stop altogether.

A century ago, science was shaken by two **BIG THEORIES.** One was **Relativity,** the other was **Quantum Theory.** Quantum Theory is about how things smaller than atoms behave in a different way from larger things. Weird quantum effects made lasers possible.

Time Travelers

**Could people ever travel through time?
Find out later—or maybe earlier?**

The Time Machine

A famous story about time travel was
H. G. Wells' *The Time Machine*, a book
written in 1895 and made into a movie in
1960. Wells imagined a future in which
beautiful humans called Eloi live a life
of leisure above ground while ugly
Morlocks slave underground.

TRUE TIME FACTS

☛ An argument
against time
travel asks what
would happen
if you traveled
back in time and killed your
grandparents before your
parents were born. Neither
your parents nor you could
have been born. But then,
who killed your grandparents?

☛ Einstein thought time
travel was impossible. He
reasoned that you cannot move
through time because you
would have to travel faster
than light. If you traveled as
fast as light, time would stop
and you would not be alive.
Some scientists disagree, but
nobody has proved him wrong!

The Tipler Cylinder

American astronomer Frank Tipler thinks you could build a time machine with a piece of super-dense material—that is, material about 10 times the mass of the Sun. You simply roll it into a straw shape a few billion miles long, then set it spinning at a few billion revolutions a minute. Then once it's spinning fast enough, it will bend space-time around it and you can send your spaceship on a spiral course along its walls. It should emerge almost instantly thousands or even billions of years away. It's pretty easy then!

Most scientific views of time travel say it might be possible to travel into the future—but not back into the past. If this is true, it explains why we've never been visited by time travelers from the future.

Bending Time

✏ Einstein showed that gravity bends space-time, so a time machine might work by using gravity to bend space-time.

✏ Time travel could be achieved using black holes—places where gravity is so powerful that it sucks in everything, including light.

✏ Black holes could be the entrances to shortcuts through space-time called wormholes.

Some scientists suggest that for each moment there are many alternative possible futures—and all of these futures actually happen. This is called the **Many Worlds Theory.**

Mad Scientists

Some of the world's most brilliant minds have also been the craziest.

Sir Isaac Newton

Newton's behavior became very strange toward the end of his life. It may be that poisoning from the many experiments he carried out with mercury had driven him a little mad.

Newton's Dog

Sir Isaac Newton loved his dog Diamond. But once Diamond knocked over a candle and started a fire that destroyed 20 years' worth of Newton's work. Newton was said to have said, "O Diamond, Diamond, thou little knowest what mischief thou hast done."

Charles Darwin

☞ The great naturalist Charles Darwin kept pet barnacles—10,000 of them!

☞ As a young man, Darwin was taught how to preserve dead animals by the freed slave John Edmonstone, who inspired his deep love of nature.

☞ After Albert Einstein's baby sister was born, his mother told him the baby would be nice to play with. After looking at it for a while, he complained, "Yes, but where are its wheels?"

☞ Andrew Crosse (1784–1855) was the inspiration for the mad fictional scientist Frankenstein. He passed an electrical current through a chemical solution as part of an experiment and living mites appeared in it. A false rumor spread that Crosse had created life.

☞ Swiss alchemist Paracelsus (1493–1541) claimed to have created a tiny artificial human called a homunculus.

☞ On his deathbed, alchemist Heinrich Agrippa (1486–1535) is said to have created an evil black demon dog that inspired the Grim in the *Harry Potter* books.

☞ Josef Papp was a Canadian–Hungarian engineer who in the 1960s claimed to have built an atomic submarine in a friend's garage and sailed it across the Atlantic Ocean.

☞ The brilliant physicist Richard Feynman used a bar as his science office, writing complex physics equations on the bar mats. As a child, he was a slow developer and couldn't talk till he was three.

Biologist Richard Owen, who coined the word dinosaur, was very keen on cutting up dead bodies. He was once desperate to get the head of a dead criminal. He bribed a jailor to get the head, but slipped and dropped it on the way home.

Henry Cavendish

☞ The great scientist Henry Cavendish dressed in clothes that were at least a century out of date. That's a little like girls today going around in ankle-length dresses, and wearing bloomers and whalebone corsets.

☞ Cavendish kept his library of books 4 miles from his home so he wouldn't be troubled by people turning up to borrow any.

That Eureka Moment

Many of the best ideas in science have come in one sudden flash of inspiration.

Bang-on

☛ Seeing this map of microwave radiation across space, scientist George Smoot said, "It's like seeing the face of God," because the pattern shows that the Universe began in a Big Bang.

TRUE YES!! FACTS

☛ In 1895, Wilhelm Röntgen detected a glow coming from a light tube sealed within a black box. He had discovered X-rays, which can shine right through things. Later, he shone X-rays through his wife's hand to make a photo of the bones.

☛ William Harvey's realization in the 1620s that the heart circulates blood around the body was probably inspired by seeing a water pump, then newly invented, in action.

☛ When Arno Penzias and Robert Wilson first detected microwave radiation coming from all over space, the echo of the Big Bang, they were convinced the signal was simply pigeon poop interfering with their antenna.

Legendary Inspiration

✎ Legend has it that Archimedes worked out the theory of buoyancy while in the bath and was so excited he ran out into the streets naked shouting, "Eureka!" This is Greek for "I've got it!"

✎ Legend also has it that Isaac Newton hit upon his theory of gravity when he was sitting under a tree and an apple fell on his head.

It was once thought that there was just one galaxy in the Universe. But one night in 1923, Edwin Hubble realized Andromeda was so staggeringly far away it could only be an entirely separate galaxy. We now know that there are billions of different galaxies.

Dreamers

☞ German scientist August Kekulé made his great break-through discovery of the ring shape of the benzene molecule in 1865 after daydreaming of a snake eating its own tail.

☞ It is said that, in February 1869, Russian chemist Dmitri Mendeleev slumped asleep at his desk and started to dream about tables— and when he woke up he realized how all the chemical elements could be organized in a table by their atomic weight.

Einstein thought of his Theory of Relativity by looking in a mirror and wondering if his reflection would vanish if he was traveling faster than light.

I Can't Believe You Thought That!

Even the most brilliant minds can sometimes get it very, very wrong.

In 1895, Lord Kelvin, the president of the Royal Society, said, "Heavier-than-air flying machines are impossible." A generation later, biplanes, such as this one, were becoming a common sight.

Way Back When

✎ In 1650, the definitive calculation showed the world began on the night before October 23, 4004 B.C.

✎ In the 1700s, scientists thought that it was a mystery substance called phlogiston that goes up in smoke when things burn.

✎ Three hundred years ago people believed that muscles contained gunpowder, which exploded to make them move.

TRUE QUOTE FACTS

☛ "The telephone has too many shortcomings to be seriously considered as a means of communication. The device is inherently of no value to us." (Western Union internal memo, 1876)

☛ "The wireless music box has no imaginable commercial value. Who would pay for a message sent to nobody in particular?" (David Sarnoff's associates, in response to his suggestion that they invest in radio in the 1920s)

☛ "Who the hell wants to hear actors talk?" (Harry M. Warner, Warner Brothers, 1927, about the prospect of movies with sound)

☛ "Spam will be a thing of the past in two years' time." (Bill Gates, head of Microsoft, speaking in 2004)

Speaking about Robert Goddard, who invented the rocket, the *New York Times* said that he "seems to lack the basic knowledge ladled out daily in high schools."

The Philosopher's Stone

☞ For thousands of years, alchemists around the world, from ancient Egypt to China, searched for ways to turn cheap metals, such as lead, into precious metals such as gold.

☞ In the 8th century A.D., the Arab alchemist Jabir ibn Hayyan said that you could turn other metals into gold by adding a red powder made from a special stone. This stone came to be known as the "philosopher's stone," and many people dedicated their whole lives to looking for it.

☞ In 1980, nuclear scientist Glenn Seaborg really did change lead to gold in a nuclear reactor.

Computers

➤➤ "I think there is a world market for maybe five computers."
(Chairman of IBM, 1943)

➤➤ "There is no reason for any individual to have a computer in their home."
(Chairman of DEC, 1977)

Numbing Numbers

We've come up with a lot of different kinds of numbers since we first started counting on our fingers.

Ancient Mathematicians

☞ When Greek mathematician Thales discovered that any triangle in a semicircle is a right-angled triangle, he celebrated by sacrificing a bull.

☞ Archimedes devised a number scheme based on a myriad myriad (100 million) to count all the grains of sand in the Universe.

☞ Maths could be dangerous in ancient times. Hypatia of Alexandria was murdered in A.D. 415 for teaching ideas that the authorities didn't like. According to her students, "her flesh was scraped from her bones with oyster shells."

16	3	2	13
5	10	11	8
9	6	7	12
4	15	14	1

Magic Numbers

☞ The game Sudoku is based on magic squares, which probably started in China 5,000 years ago. The numbers in each row, column, or diagonal add up to the same total.

☞ After trying to work out how rabbits breed, Fibonacci discovered a series of numbers in which each is added to the one before: 1, 1, 2, 3, 5, 8, 13, 21, 34, 55, etc. The number of petals in a sunflower is always one from this "Fibonacci Series."

Big and Little Numbers

✏ Prefixes are letters that come at the front of a word. Some prefixes describe big numbers: mega (1 with 6 zeros), giga (1 with 9 zeros), tera (12 zeros), peta (15), exa (18), zetta (21), and yotta (24).

✏ The prefixes for the smallest numbers are: micro (0 with 6 zeros after the decimal point), nano (9 zeros), pico (12), femto (15), atto (18), zepto (21), and yocto (24).

☞ A standard pack of 52 playing cards can be dealt out in 806 million billion billion billion billion billion billion ways (that's 806 followed by 60 zeros).

☞ Astronomer Sir Arthur Eddington estimated that there are 15,747,724,136,275,002,577,605, 653,961,181,555,468,044,717,914, 527,116,709,366,231,425,076,185, 631,031,296 protons in the Universe. It was an ambitious thing to guess. And he was wrong.

☞ Scientists have come up with some very silly names for the biggest numbers of all, including picoboos, gigalos, terabulls, and nanogoats.

Greek geek Zeno showed how logic proved that, in a race, a hare can never catch up with a tortoise if the tortoise has a head start. This is called a paradox because we know it can't be true.

Slinky Genes

Genes are the chemical code in our cells that make us what we are.

TRUE GENE FACTS

☞ Inside every cell in every living thing is a tiny twist of a chemical called DNA. DNA is basically life's recipe book, and the DNA in each of your body's cells contains all the information needed to make a new you.

☞ DNA contains instructions for every single ingredient of you. These instructions are called genes.

☞ Your body's cells have each got about 23,000 genes in them. But before you get too smug about how complex you are, even a nematode worm's got 20,000 genes. And a mustard plant's got 27,000.

A woman's claims to be Anastasia, the "lost" Russian princess whose family were killed in 1917, were proved false by comparing her DNA with some from the victims' descendants.

Genetic modification, or GM, means taking genes out of one organism's DNA and putting them in another. In 1985, scientists put genes for human growth factor in a pig, which then grew so big it became crippled with arthritis.

Some DNA, called mitochondrial DNA, changes very little over time. It has been used to trace the ancestry of the entire human race. The evidence suggests that we are all descended from a woman who lived in Africa 250,000 years ago.

→ Scientists could genetically alter spinach to make it taste like chocolate so children eat it. They also hope to take out the gene for the chemical in onions that makes you cry—and make no-tears onions.

→ French scientists gave jellyfish genes to a rabbit, making it glow in the dark.

Twisty DNA

In 1953, Rosalind Franklin, Francis Crick, Maurice Wilson, and James Watson found that DNA is shaped like a twisted rope ladder, a shape called a double helix.

Suicidal Potatoes

In the mid 1990s, scientists came up with a clever way to stop fungus spreading through a potato crop. They inserted the gene for a substance called barnase into potato plants. Now the plants produce floods of barnase whenever they are attacked by fungus. The barnase kills the plants. The infected plant effectively commits suicide, and this stops the fungus from spreading.

"What could be better than a self-shearing sheep?" thought Australian scientists. They put genes into a sheep that made its wool drop out once it reached a certain length. It certainly saved on the shearing. But the poor naked sheep suffered terrible sunburn.

Essential Facts

Amazing Body

☛ Heaviest organ: skin, 5–9 pounds.

☛ Largest cell: megakaryocyte, in the bone marrow, 0.2 mm across.

☛ Smallest cell: neuron, in the brain, 0.005 mm across.

☛ Strongest joint: hip.

☛ Biggest muscle: gluteus maximus, in the buttocks.

☛ Smallest muscle: stapedius, in the inner ear.

☛ Longest muscle: sartorius, in the inner thigh.

☛ Widest muscle: external oblique, running around the upper body.

☛ Largest blood vessel: aorta, about the size of a garden hose.

☛ Loudest burp ever: 118.1 decibels.

☛ Longest beard ever: 16 feet, belonging to Norwegian Hans Langseth.

☛ Longest ever nails: combined length of 23 feet, belonging to Sridhar Chillal of India.

☛ Oldest person ever, whose age was proven: Jeanne Calment, a woman from France, who died in 1997 aged 122.

☛ Oldest man ever, whose age was proven: Dane Christian Mortensen, who died in 1998 at the age of 115.

Sickness and Health

☛ First ever skin grafts: India in the 6th century B.C.

☛ First successful amputation under anesthetic: London, 1847.

☛ Deadliest outbreak of disease ever: the Black Death, which killed 25 million people between 1347 and 1351.

☛ Fastest amputation: by British surgeon Robert Liston, who could saw off a leg in 28 seconds.

☛ Most common disease: cold.

☛ Deadliest disease that is curable with treatment: tuberculosis.

☛ Most infectious disease: measles.

☛ First successful kidney transplant: by R.H. Lawler, in Chicago, Illinois, in 1950.

☛ First successful heart transplant: by Christiaan Barnard in Cape Town, South Africa, in 1967.

☛ First successful heart and lung transplant: by Dr Bruce Reitz of Stanford, USA, in 1981.

☛ First hand transplant: by Frenchman Jean-Michel Dubernard, in 1998.

☛ First bionic limb: arm fitted to Jesse Sullivan of the United States in 2002.

They Did What?!

☞ Maddest Roman emperor: Caligula (A.D. 12–41) who wandered his palace at night, commanding the sun to rise.

☞ Longest reigning king: Pepi II of Egypt (2275–2175 B.C.), 94 years.

☞ Only teenage girl ever to command an army: Joan of Arc, the French against the English at Orléans, in 1429.

☞ First criminal to be captured by Morse Code: Dr. Crippen in 1910, when a message was sent to the ship on which he was escaping.

☞ Most prolific serial killer: British doctor Harold Shipman killed 500 patients.

☞ Richest person in the world: Bill Gates, founder of Microsoft, worth $60 billion.

☞ First man to run 100 metres in under 10 seconds: Jim Hines of the United States in 1968.

☞ First man to run a mile in under four minutes: British athlete Roger Bannister, in 1954.

☞ Youngest world number one golfer ever: American Tiger Woods, aged 21.

☞ Highest IQ (intelligence quotient) ever recorded: American Marilyn vos Savant's 230.

The Human World

☞ First ever skyscraper: Home Insurance Building, Chicago, in 1885.

☞ Tallest building in the world: Taipei 101, in Taiwan, 1,500 ft high. From 2009, it will be the 2,000-ft-high Burj Dubai tower in Dubai.

☞ Biggest city: Tokyo, 35.5 m people.

☞ First ever patent: given to Filippo Brunelleschi in Florence, in 1421, for a type of barge.

☞ First modern flushing toilet: Castle Ehrenburg, Germany, in 1860, for the exclusive use of Queen Victoria.

☞ First to invent antibiotic drugs: Alexander Fleming in 1928, penicillin.

☞ First powered flight: brothers Orville and Wilbur Wright on 17 December 1903 in North Carolina, USA.

☞ World's first Maglev train service: from Shanghai Airport to Shanghai city centre, China, opened in 2004.

☞ First ever mass-produced popular car: Ford Model T, in 1905.

☞ Best-selling car of all time: VW Beetle, which sold 20 million.

☞ Most powerful computer in the world: Blue Gene/L, which has a calculating speed of 360 tetraflops (one tetraflop means one trillion calculations per second).

Earth and Space

☛ Average surface temperature of the world's oceans: 63 °F.

☛ Biggest desert in the world: Sahara.

☛ First people to climb Everest and return alive: Edmund Hilary and Tensing Norgay on 29 May 1953.

☛ Shortest river: North Fork Roe River in Montana, USA, 54 feet long.

☛ Most active volcano: Kilauea, on Hawaii, continuously active since 1983.

☛ Biggest volcano: Mauna Loa, on Hawaii, 18,000 cubic miles.

☛ Hardest organisms: extremophiles, which may also live on Mars.

☛ Largest planet in the Solar System: Jupiter, 1321 times the volume of Earth.

☛ Smallest planet: Mercury, 5 percent the volume of Earth.

☛ Coldest place in the Solar System: Neptune's moon Triton, with surface temperatures of -393 °F.

☛ Closest star to Earth: Proxima Centauri, 4.22 light years away.

☛ Brightest star in the night sky: Sirius, which is 8.6 light years away.

☛ Number of men to walk on the Moon: 12, all between 1969 and 1971. The first was Neil Armstrong, followed a few minutes later by Buzz Aldrin.

☛ Most spectacular meteor shower: the Leonids, which occur every 33 years when a comet passes close to Earth.

Crazy Nature

☛ USA's deadliest tornado: Tri-State, in 1925, which killed 695 people.

☛ Deadliest storm of the 20th century: Bangladesh, 1970, over 250,000 killed.

☛ Strongest recorded gust of wind: 231 mph on Mount Washington, New Hampshire, in 1934.

☛ Biggest clouds: cumulonimbus, up to 12 miles tall.

☛ Chances of two snowflakes having the same pattern: 10^{158} to 1 against.

☛ Coldest temperature ever recorded: -128 °F in Antarctica.

☛ Biggest forest: 6.5 million square miles, in Siberia.

☛ Oldest living individual tree: Methuselah, a bristlecone pine in California, which is 4700 years old.

☛ Smallest flowering plant: *Wolffia arrhiza*, less than 1 mm across.

☛ Biggest meat-eating plant: nepenthes, whose animal traps can grow over one foot long.

☛ Biggest living organism: a fungus in Oregon, USA, which lives underground and spreads over 2,200 acres.

Awesome Animals

☞ Heaviest living bird: ostrich, 300 pounds.

☞ Heaviest bird that can fly: great buzzard, 40 pounds.

☞ Smallest mammal: the bumblebee bat, which weighs $\frac{1}{10}$ ounce.

☞ Fastest land animal: cheetah, which can run at 70 mph.

☞ Largest swarm of locusts ever recorded: 10 billion, in Kenya in 1954.

☞ Oldest species of fish: coelacanth, 350 million years old.

☞ Largest and longest-lived turtle: leatherback, can live for 170 years.

☞ Biggest reptile: saltwater crocodile, up to 15 feet long.

☞ Smallest reptile: dwarf gecko, less than one inch long.

☞ Largest animal: blue whale, up to 100 feet long and 200 tons in weight.

☞ Largest land animal: African elephant, which can weigh up to 20,000 pounds.

☞ Only poisonous mammal: duck-billed platypus, with venomous spurs on its hind feet.

☞ Longest jumper for its size: southern cricket frog, which can leap 6 feet: that's 60 times its 1-inch-long body.

Weird Science

☞ Smallest known particle: quark.

☞ Normal terminal velocity of a skydiver: 121 mph.

☞ Terminal velocity if skydivers pull in their limbs: 200 mph.

☞ Strength of the Moon's gravitational pull: 17 percent that of Earth.

☞ Smallest Standard International (SI) unit of weight: yoctogram, used to weigh subatomic particles.

☞ Largest SI unit of weight: yottagram, which is 1,000,000,000,000,000,000, 000,000 grams.

☞ Amount of energy turned to light by an ordinary light bulb: about 5 percent.

☞ Energy per day needed by the average adult human: 8,000 kilojoules.

☞ Speed of light: 186,282 miles per second.

☞ First ever story about time travel: Samuel Madden's 1733 book *Memoirs of the Twentieth Century*.

☞ Longest continuous laboratory experiment: at University of Queensland in Australia, where pitch has been pouring down a hole since 1930 at a rate of one drop every ten years.

☞ Largest known prime number: $2^{32,582,657}-1$, discovered in 2006 by Curtis Cooper and Steven Boone.

Index

Acknowledgments

All artwork supplied by Guy Harvey

Photo credits:
b – bottom, t – top, r – right, l – left, m – middle
Cover: l Dreamstime.com/Steve Luker, m NASA, r Buddy Mays/Corbis
1 Drazen Vukelic/Dreamstime.com, 3 Dreamstime.com, 4 Dreamstime.com/Joao estevao Andrade de freitas, 6-7 Andrew Davis/Dreamstime.com, 6t Wayne Abraham/Dreamstime.com, 6b Dreamstime.com/Kathy Wynn, 7t Dreamstime.com, 7b John Kounadeas/Dreamstime.com, 8-9 Dreamstime.com/Monika Wisniewska, 9bl Dreamstime.com/Steve Luker, 9bm Dreamstime.com/Janet Carr, 9br Dreamstime.com, 10 Dreamstime.com, 11t Corbis/Roger Ressmeyer, 11b Dreamstime.com/James Steidl, 12 Dreamstime.com/Steve Luker, 13t Dreamstime.com/Fallenangel, 13b Dreamstime.com/Andy Piatt, 14 Corbis/Bob Sacha, 15l C/Mediscan, 15t Dreamstime.com/Graça Victoria, 16 Dreamstime.com/Noriko Cooper, 17t Dreamstime.com/Mandy Godbehear, 17b Dreamstime.com/Silas Brown, 18t Dreamstime.com, 18b Grace/zefa/Corbis, 19 C/Larry Williams, 20-21 Dreamstime.com/Janet Carr, 20t Dreamstime.com/Jason Stitt, 21t Dreamstime.com/Peter Galbraith, 22 Dreamstime.com/Showface, 23t Visuals Unlimited/Corbis, 23b Dreamstime.com/Eraxion, 24-25 Dreamstime.com/Sebastian Kaulitzki, 24 Corbis/Angelo Christo/zefa, 25 Corbis/Howard Sochurek, 26 Dreamstime.com/Milan Kopok, 27t Dreamstime.com/Daniel Schmid, 27b Dreamstime.com/Jeecis, 28 Heiko Wolfraum/dpa/Corbis, 29b Dreamstime.com, 29t Lester V. Bergman/CORBIS, 30b Dreamstime.com/Simone Van Den Berg, 30t Dreamstime.com/Monika Wisniewska, 31 Lester V. Bergman/Corbis, 32 Dreamstime.com/Andrew Taylor, 33t Dreamstime.com/Sherrie Smith, 33b Parrot Pascal/Corbis Sygma, 34-35 Dreamstime.com/Scott Rothstein, 35l Dreamstime.com/Vladimirs, 35m Dreamstime.com/Lee Reitz, 35r Dreamstime.com, 36 Dreamstime.com, 37 Reuters/Corbis, 38b Dreamstime.com/Lee Reitz, 38t Dreamstime.com/Andra Cerar, 39 Bettmann/Corbis, 40t Bettmann/Corbis, 40b Dreamstime.com/Eldoronki, 41l Richard A. Cooke/Corbis, 41r Dreamstime.com, 42 Bettmann/Corbis, 43t Dreamstime.com/Vladimirs, 43b Bettmann/Corbis, 44b Dreamstime.com/Alex Kalmbach, 44t Jon Feingersh/zefa/Corbis, 45t Dreamstime.com/Yuen Che Chia, 45b Dreamstime.com, 46-47 Dreamstime.com, 46l Dreamstime.com/Ewa Walicka, 46r Dreamstime.com/Ryan Jorgensen, 47 Dreamstime.com/Simone van den Berg, 48-49 Dreamstime.com/Kati Neudert, 49t Visuals Unlimited/Corbis, 49b Dreamstime.com, 50l Dreamstime.com, 50b Dreamstime.com/Bruce Macqueen, 51 Dr. Milton Reisch/Corbis, 52t Dreamstime.com, 52b Anthony Bannister; Gallo Images/Corbis, 53t Dreamstime.com/Scott Rothstein, 53b Dreamstime.com/Edward Westmacott, 54b Dreamstime.com/Ljupco Smokovski, 54t Dreamstime.com/Fallenangel, 55 Deamstime.com, 56-57 Bettmann/Corbis, 56t Dreamstime.com/Clint Scholz, 57t Pascal Rossignol/Reuters/Corbis, 58 Ed Kashi/Corbis, 59 Robert Galbraith/Reuters/Corbis, 60-61 Dreamstime.com, 62l Dreamstime.com/Antonio Ballesteros, 61m Dreamstime.com/Aleksandr Lobanov, 61r Dreamstime.com/Ryan Jones, 62t Dreamstime.com/Antonio Ballesteros, 62b Bettmann/Corbis, 63l d8/Scott Rothstein, 63r d8/Lori Martin, 64 Dreamstime.com/Eti Swinford, 65l Louise Gubb/Corbis, 65r Dreamstime.com/James Hearn, 67l Stefano Bianchetti/Corbis, 67r Handke-Neu/Corbis, 68t Dreamstime.com/Rafael Laguillo, 68b Dreamstime.com/Vladimir Pomortsev, 69l Richard Melloul/Sygma/Corbis, 69r Michael Nicholson/Corbis, 70-71 Dreamstime.com/Romulus Hossu, 70b Dreamstime.com/Ewa Walicka, 71b Hulton-Deutsch Collection/Corbis, 72b Dreamstime.com/Aleksandr Lobanov, 73 Siemonet Ronald/Corbis Sygma, 74t Bettmann/Corbis, 74b Dreamstime.com/Simon Gurney, 75t Dreamstime.com/Michael Shake, 75b Dreamstime.com/Scott Rothstein, 76b Dreamstime.com/Lyn Baxter, 76t Dreamstime.com, 77 John Bryson/Sygma/Corbis, 78t Dreamstime.com, 79t Group of Survivors/Corbis, 79b Dreamstime.com/Olaf Schlueter, 80 Dreamstime.com/Vladimir Pomortsev, 81m Hulton-Deutsch Collection/Corbis, 81b Dreamstime.com/Ryan Jones, 82t Corbis, 82b Dreamstime.com, 83t Dreamstime.com/Sandra Henderson, 83b Reuters/Corbis, 84t Bettmann/Corbis, 84b Dreamstime.com/Steve Degenhardt, 85t Dreamstime.com, 85b Dreamstime.com/Dragan Trifunovic, 86-87 Ralph Paprzycki/Dreamstime.com, 87l Andrew Davis/ Dreamstime.com, 87m Shuttlecock/Dreamstime.com, 87r 88 Dreamstime.com, 89l Dreamstime.com/Utsav Arora, 89r Ralph Paprzycki/Dreamstime.com, 90t Andrew Davis/ Dreamstime.com, 90t Jonathan Blair/Corbis, 92t Maurice Nimmo; Frank Lane Picture Agency/Corbis, 92-93 Utsav Arora/Dreamstime.com, 93t Dreamstime.com/Kirill Zdorov, 93b NASA, 94 Bettmann/Corbis, 95t Edite Artmann/Dreamstime.com, 95b Dreamstime.com, 96t Anthony Hall/Dreamstime.com, 97 Gideon Mendel/Corbis, 98l Joe Gough/Dreamstime.com, 98r Dreamstime.com/Soldeandalucia, 99t Johanna Goodyear/Dreamstime.com, 99b Leon Forado/Dreamstime.com, 100l Editoria/Dreamstime.com, 100r Dreamstime.com, 101l Dreamstime.com, 101r Corbis, 102b Joris Van Den Heuvel/Dreamstime.com, 103l Louie Psihoyos/Corbis, 103r Pablo Eder/Dreamstime.com, 104-105 Rob Bouwman/Dreamstime.com, 104 Robert Creigh/Dreamstime.com, 105t Lein De Leon/Dreamstime.com, 106-107t Wayne Abraham/Dreamstime.com, 106-107b Wayne Mckown/Dreamstime.com, 107r Prestong/Dreamstime.com, 108 Bettmann/Corbis, 109t Ian Klein/Dreamstime.com, 109b John Leung/Dreamstime.com, 110l Linda & Colin Mckie/Dreamstime.com, 110-111 Shuttlecock/Dreamstime.com, 111 Hulton-Deutsch Collection/Corbis, 112-113 NASA, 113l NASA, 113m Digital Vision, 113r Roger Degen/Dreamstime.com, 114-115 Naluka/Dreamstime.com, 114 John Kounadeas/Dreamstime.com, 115t NASA, 115b Wikipedia/msauder, 116-117 Roger Degen/Dreamstime.com, 116t Dreamstime.com/Vladimir Pomortsev, 117b Laurin Rinder/Dreamstime.com, 118t Dreamstime.com/Jose Fuente, 118b Didrik Johnck/Corbis, 119t Dreamstime.com, 120-121 Nicole Andersen/Dreamstime.com, 121 Isospin123/Dreamstime.com, 122 Chris 73/ http://commons.wikimedia.org/wiki/Image:Bridge_across_continents_iceland.jpg, 123 Digital Vision, 124t ESA, 124b NASA, 125 all NASA, 126-127 all NASA, 128-129 NASA, 128 Mark Bond/Dreamstime.com, 129t NASA, 129b ESA/DLR/FU Berlin (G. Neukum), 130-131 NASA, 132-133 NASA, 132 Corbis, 133 Sebastian Kaulitzki/Dreamstime.com, 134l Bettmann/Corbis, 134-135 NASA, 135t Bettmann/Corbis, 135b NASA, 137t Hyside/Dreamstime.com, 137b NASA, 138-139 Ian Francis/Dreamstime.com, 139l Tiger Darsn/Dreamstime.com, 139m Dreamstime.com, 139r 140t Bettmann/Corbis, 140b Dreamstime.com, 141 Dreamstime.com, 142-143 Eric Nguyen/Corbis, 142t Amelia Takacs/Dreamstime.com, 143 Michael Freeman/Corbis, 144 Amy Ford/Dreamstime.com, 145 Stringer/USA/Reuters/Corbis, 146-147 Lanceb/Dreamstime.com, 147t James Robert Fuller, 147b Andy Nowack/Dreamstime.com, 148 Dreamstime.com/Marcelo Zagal, 149t Jim Reed/Corbis, 149b Jerry Horn/Dreamstime.com, 150-151 Matej Krajcovic/Dreamstime.com, 150t Dreamstime.com/Jan Will, 151r Dean Conger/Corbis, 151-152 Tiger Darsn/Dreamstime.com, 152l Gail Johnson/Dreamstime.com, 153 Andreasguskos/Dreamstime.com, 154 Dreamstime.com, 155t Elena Elisseeva/Dreamstime.com, 155b Dreamstime.com/Alantduffy1970, 156b Dreamstime.com, 156-157 Kar Yan Mak, 157r Dreamstime.com, 158l Neil Miller; Papilio/Corbis, 158t Norman Chan/Dreamstime.com, 160b Lester V. Bergman/Corbis, 160r Sergey Zholudov/Dreamstime.com, 161r Adam Płonka/Dreamstime.com, 162l Oktay Ortakcioglu/Dreamstime.com, 162t Anette Linnea Rasmussen/Dreamstime.com, 163t Nik Wheeler/Corbis, 163b Alexander Kolomietz/Dreamstime.com, 164-165 Corbis, 165l 172t Dreamstime.com/Joao estevao Andrade de freitas, 165m Dreamstime.com/Stephen Girimont, 165r Dreamstime.com/Andy Heyward, 166 Dreamstime.com/Kathy Wynn, 167b Paul Wolf/Dreamstime.com, 168t Dreamstime.com, 168b Dreamstime.com/David Hancock, 169 Buddy Mays/Corbis, 170t Dreamstime.com/Christopher Marin, 170b Dreamstime.com/Anthony Hathaway, 171 Dreamstime.com/David Pruter, 172t Dreamstime.com/Joao estevao Andrade de freitas, 172b Dreamstime.com/Marek Kosmal, 173t Dreamstime.com/Ron Brancato, 173b Dreamstime.com/Chris Fourie, 174-175 Stuart Westmorland/Corbis, 174l Amos Nachoum/Corbis, 175t Dreamstime.com/Asther Lau Choon Siew, 176l Dreamstime.com/Isabel Poulin, 176r Dreamstime.com/Stephen Girimont 177t Eendicott/Dreamstime.com, 177b Dreamstime.com/Ken Cole, 178l Dreamstime.com/Can Balcioglu, 178t Joe McDonald/Corbis, 179 Dreamstime.com/Andy Heyward, 180 Eric Coia/Dreamstime.com, 181b Dreamstime.com/Rachel Barton, 182 Tim Davis/Corbis, 182-183b Dreamstime.com/Vladimir Kindrachov, 184t John Loader/Dreamstime.com, 184b Dreamstime.com/Brett Atkins, 185b Mono Andes, 186 Dreamstime.com/Stephen Inglis, 187l Dreamstime.com/Fred Goldstein, 187b Jeff Clow/Dreamstime.com, 188 Denis Scott/Corbis, 189t Bernard Breton/Dreamstime.com, 189b Andre Nantel/Dreamstime.com, 191l Matthias Kulka/zefa/Corbis, 191m Dreamstime.com/Andreus, 191r Dreamstime.com/Anita Patterson Peppers, 192t Sonya Etchison/Dreamstime.com, 192b NASA, 193 Drazen Vukelic/Dreamstime.com, 194l Dreamstime.com, 194t Dreamstime.com, 195t Rainer/Dreamstime.com, 195b Dreamstime.com, 196-197 Jacek Kutyba/Dreamstime.com, 196t Steve Lupton/Corbis, 197t Red2000/Dreamstime.com, 197b Matthias Kulka/zefa/Corbis, 198t NASA, 198b Stefan Baum/Dreamstime.com, 199 Dreamstime.com/Andreus, 200-201 Bob Sacha/Corbis, 201t Dreamstime.com/Anita Patterson Peppers, 201b Peter Kim/Dreamstime.com, 202l Dreamstime.com, 203t Kevin Fleming/Corbis, 204-205 Dreamstime.com, 204l Bettmann/Corbis, 206b Dreamstime.com, 207t Bettmann/Corbis, 208l Dreamstime/Tyler Olson, 208t NASA, 208-209 Esa/V. Beckmann, 209 Bettmann/Corbis, 210l Dreamstime/Martin Plsek, 211t NASA, 211b NASA, 212t Dreamstime.com/Laura Bulau, 213 James Steidl/Dreamstime.com, 214t Rykoff Collection/Corbis, 214b Anthony Gaudio/Dreamstime.com, 214-215 Matthias Kulka/zefa/Corbis, 215 Jiri Castka/Dreamstime.com